DATE DUE

MY 2 6 '00			

FORENSIC ENGINEERING

FORENSIC ENGINEERING
ENVIRONMENTAL CASE HISTORIES
FOR CIVIL ENGINEERS AND GEOLOGISTS

Gerard Shuirman
James E. Slosson

ACADEMIC PRESS, INC.
Harcourt Brace Jovanovich, Publishers

San Diego   New York   Boston   London   Sydney   Tokyo   Toronto

Cover photograph courtesy of John C. McCarthy, Esq. (see page 176 for details).

This book is printed on acid-free paper. ∞

Academic Press, Inc.
1250 Sixth Avenue, San Diego, California 92101-4311

United Kingdom Edition published by
Academic Press Limited
24–28 Oval Road, London NW1 7DX

Library of Congress Cataloging-in-Publication Data

Shuirman, Gerard.
    Forensic engineering : environmental case histories for civil
engineers and geologists / Gerard Shuirman, James E. Slosson.
        p.   cm.
    Includes bibliographical references and index.
    ISBN 0-12-640740-1
    1. Forensic engineering--Environmental aspects.   2. Forensic
engineering--Case studies.   I. Slosson, James E., date.
II. Title.
TA219.S58   1992
624--dc20   91-34361
                                                              CIP

PRINTED IN THE UNITED STATES OF AMERICA

92 93 94 95 96 97   QW   9 8 7 6 5 4 3 2 1

This book is dedicated to our wives Shirl Armstrong-Shuirman and Nancy Slosson for their suggestions and loving support. It was Shirl, a former literary agent, who started it all. After listening to some of our more bizarre forensic experiences over dinner one evening, she suddenly exclaimed, "You guys have a book just waiting to be written!"

The book is also dedicated to the memory of our friend Max Truex, an Olympian in fact and spirit. A skilled trial attorney, Max introduced both authors to the forensic challenge.

CONTENTS

FOREWORD

Increasingly architects, engineers, geologists, and landscape architects are called upon to be expert witnesses in liability suits resulting from flood, erosion, landslide, mudslide, or other types of natural hazard–related damage. They may be called upon to defend their actions or those of their agency or organization. Or, they may act as independent experts for individuals and their attorneys who are either bringing suits or defending themselves against such suits.

Many of these liability suits are between adjacent private landowners. Others are between private landowners and government entities. In either instance, the individual bringing the suit claims that the party being sued has caused or increased the natural hazard damage.

Although the specifics of suits differ (usually based upon a claim of "negligence"), the expert in such a suit is usually called upon to establish: (*a*) that the party being sued acted unreasonably in the circumstances, and (*b*) that this unreasonable action caused the damage. Reasonableness depends on the nature of the activity, the foreseeability of the damage, the potential seriousness of the damage, the standard of conduct applied generally in the profession and in the specific region, and other factors.

The role of the expert witness is not an easy one for the uninitiated. One must be prepared to deal with the complexities of proof and the inevitable confrontations with and challenges by opposing attorneys and experts.

This book will be of great value to those called upon or wishing to volunteer as expert witnesses; it fills a void in the literature. The book clearly explains the needs of an expert, the relationship of the expert to the client and to the attorney, challenges one may face, and the proper orientation one needs as an expert. Through a variety of case studies, it illustrates investigative techniques, case preparation, and data presentation to prove reasonableness or unreasonableness of conduct and causation. This book could only have been written by individuals highly experienced in the courtroom as well as in the field.

The book does something more, however, for which I commend the authors. It recognizes that experts have a responsibility not only to their clients but also to the court and to the judicial process. It recommends integrity and honesty—essential values if suits are to result in justice, a sometimes deemphasized goal. Integrity and honesty combined with good preparation and some judicial smarts also win lawsuits.

I commend the authors for writing this book. You will find that it provides fascinating insight into the world of the expert even if you never appear in court.

Jon Kusler

Jon Kusler has been an attorney and policy analyst for 24 years. He specializes in hazard, land use law, and water use law. He is Chief Counsel and Research Director for the Association of State Floodplain Managers and author of numerous reports for the American Bar Association's Special Committee on Housing and Urban Development, Federal Emergency Management Agency, and Natural Hazards Research and Applications Information Center.

PREFACE

Webster's New World Dictionary defines the word *forensic,* in part, as "of or suitable for public debate." This definition is an apt and succinct description of our basic reason for writing this book.

Most of our professional activity for the past 15 years has been of a forensic nature, culminating in our becoming expert witnesses in court cases. Dr. Slosson, in addition to his consulting practice, has spent a considerable portion of his career working with code and regulatory commissions, as a consultant employed for second party reviews, as the State Geologist for the State of California, and as a professor of geology. Before becoming involved in forensic work, Mr. Shuirman was president of two civil engineering consulting companies specializing in the planning and design of a wide variety of public and private projects. In our positions we have seen an alarming increase in the number of environmentally related failures and disasters that end up in litigation. As a result, we feel compelled to share our experiences and thoughts regarding not only the forensic process but also broader questions of professional, political, and societal concerns related to reversing, or at least slowing, this costly and resource-wasting trend.

Much of the previous literature and special course material dealing with forensic engineering emphasizes the procedures involved in the typical forensic investigation along with providing instructions for how to be an effective expert witness during litigation. We also address these subjects from our own perspectives. However, our major concerns are the ethical and professional aspects of the entire forensic process and the "post mortem" analyses of specific environmentally related failures and disasters: Why they occured, how they could have been prevented, and what lessons should have been learned. The following questions, among others, are indeed suitable for public debate.

1. Should an engineer or geologist employed in a forensic assignment ever, under any circumstance, play the role of an advocate? (In our experience, most do.)
2. What can be done to ensure that the courts receive technical evidence of the highest possible quality upon which to base their decisions? (In our opinion, this often does not happen. Many unwise legal precedents and

laws related to environmental hazards and land development are stemming from inaccurate and inadequate investigations and advocacy type court testimony on the part of expert witnesses.)

3. What can be done to prevent or restrict inappropriate land development in high-risk areas such as floodplains, earthquake fault zones, and landslide-prone hillside areas? (In some jurisdictions this problem is finally beginning to receive the attention it deserves.)

4. How can civil engineers and engineering geologists and their professional organizations exert more influence on politicians and administrators in the decision-making process regarding land use in hazardous areas? (Often the most qualified are intentionally or inadvertently left out of this process.)

5. Conversely, what can be done to prevent politicians from exerting influence on administrators and professionals to approve projects in high-risk areas without adequate investigation and mitigation measures? (Developers' political contributions carry considerable clout.)

6. What can be done to ensure that corners would not be cut on environmentally sensitive development projects in the interest of saving costs? (Our experience indicates that many projects sustain multimillion dollar damage owing to cost savings of several thousand dollars.)

7. Do the traditional methods of evaluating alternative designs and preparing cost–benefit analyses need to be updated? (In our opinion, the potential costs related to litigation are so significant that they cannot be disregarded merely because they are less predictable than other costs.)

8. What policies and procedures should be adopted to ensure better quality control in design and construction of land development projects in hazardous areas? (This is especially important in an era of dwindling tax dollars available to public entities responsible for reviewing and monitoring such projects.)

9. Once a failure or disaster has occurred, what can be done to emphasize mitigation as a solution rather than litigation? (We have been involved in failures where the estimated cost of mitigation is exceeded by attorney fees.)

These questions involving broad professional and public policy are typical of those that have arisen from our forensic experiences. Implicit in several of the questions is the need for better public education and awareness:

- Lawmakers need to be better informed in order to provide adequate legislation.
- Judges need to be better informed in order to interpret laws in a reasonable way as well as to make better decisions regarding admissibility of evidence.
- The general public needs to be better informed in order to demand appropriate action from their representatives, public administrators, and

the judiciary, as well as to make better decisions at the voting booth. Perhaps we need fewer public broadcasting station series on plate tectonics and more on geologic hazards.

Because of our particular areas of expertise, the case histories are confined to environmental-type failures and hazards. They encompass flooding, land subsidence, landslides, erosion, sediment deposition, and related land movement. Despite the limited scope of the case histories in relation to all engineering type failures, the principles involved in the investigations, analyses, and adjudications apply in general to forensic engineering projects in other fields.

The case histories represent forensic assignments in which we have been directly and actively involved. Each case history describes in detail the investigation, analysis, and litigation phases. The litigation is commented upon solely from our viewpoints as expert witnesses. Each case history is concluded with a post mortem discussion of how the failure could have been avoided and what should have been learned in terms of future planning and design as well as public policy.

Our original intent was to write a book that would serve as a guide and reference for civil engineers and geologists who are either involved in or contemplating becoming involved in forensic work. Later it occurred to us that much of what we had to say might appeal to a wider spectrum of readers, such as engineers in other branches, land and environment planners, public officials and administrators, hazard insurers, and attorneys who specialize in real estate, public works, or environmental law. Now, we also envision the book as a text or reference for upper-level and graduate-level civil engineering and geology students.

To make the book more useful and interesting to a broader-based readership we incorporated the technical aspects of the case histories (predicated largely on our reports and published papers) into a more anecdotal writing style. We believe the uniqueness of the book lies not only in the writing methodology employed but also in its emphasis on the professional, ethical, and moral questions encountered in the forensic process and the thought-provoking nature of the post mortem comments following each case history.

ACKNOWLEDGMENTS

The authors are indebted to many of our colleagues and clients who have contributed in some way to the case histories contained in the book. Among those we wish to single out are Alexander R. Ball, R. G.; Steven R. Borer, Esq.; David B. Casselman, Esq.; Cynthia Cronin; Arnold K. Graham, Esq.; Douglas L. Hamilton, P. E.; John F. Krattli. Esq.; Dr. Robert C. McArthur, P. E.; John C. McCarthy, Esq.; Paul L. McClay, R. G.; Dr. Richard L. Meehan, P. E.; James C. Powers, Esq.; Michael F. Richman, Esq.; and Delmar D. Yoakum, P. E.

Special gratitude is due H. Crane Miller, an attorney who specializes in natural hazard law and legislation. Although not associated with us in any of the case histories, Mr. Miller took the time to painstakingly edit the entire manuscript and to make many valuable suggestions.

A colleague who made vital contributions to the book is Dr. Vincent S. Cronin, an Assistant Professor of Engineering Geology at the University of Wisconsin–Milwaukee. While a member of Dr. Slosson's consulting staff, Dr. Cronin wrote several insightful technical reports and papers regarding the Big Rock Mesa landslide (see Chapter 3) that formed the basis for the geologic investigation portion of that chapter.

Finally, we are deeply indebted to Gay Havens, our fine illustrator, who has a unique knack of transforming dry data into readily understandable charts and maps.

1 FORENSIC INVESTIGATIONS: A GROWTH INDUSTRY

INTRODUCTION

In the 1970s most of us associated the word *forensic* with Jack Klugman's role as a medical examiner in the television show "Quincy"—that is, one who examines and dissects corpses for evidence of medical malfeasance or foul play.

Since that time there has been a veritable explosion of forensic work encompassing many professional fields. What started with medicine has spread to other professions and disciplines, in particular to the earth and environmentally related sciences. All branches of engineering, including civil and geotechnical, have become highly involved in forensic investigations. The American Society of Civil Engineers has even established a special section for forensic specialists. Many civil engineers and geologists spend most of their professional time doing forensic work; some devote all of their time to it.

What exactly does the word forensic mean, and how does it apply to civil engineering and geology? Definitions abound, but basically, for this book, it implies the scientific investigation of accidents and environmentally related failures and disasters in order to discover their causes. These investigations can be made for many reasons, and their findings and conclusions can be used in myriad ways. A forensic investigation need not necessarily be performed for legal purposes nor result in litigation. However, most are done for legal reasons, with lawsuits usually filed in advance of any serious forensic work.

Why has the demand for forensic experts in civil engineering and geology been skyrocketing? The reasons are varied, complex, and, in many cases, interrelated. They should be carefully examined, if for no other purpose than to find ways to stem the growth. Some of the reasons are discussed here.

An Overabundance of Attorneys?

A pervasive feeling in our society is that there are too many trial lawyers chasing too few legitimate lawsuits, thereby encouraging prospective plaintiffs to sue before ever examining the merits of the litigation. With

1

contingency legal fees the would-be plaintiff has little to lose in initiating a lawsuit. The attorney has much to lose but may accept the case in spite of a lack of experience or knowledge in a particular field. However, more often the attorney will decide to gamble on a questionable or creative-type case simply because the potential rewards (generally 30–40% of the settlement or verdict), coupled with the perception of an inept defense attorney, make it worthwhile. Despite the profusion of frivolous cases, in our opinion overzealous attorneys are not the primary reason for the accelerating growth of lawsuits requiring forensic services from civil engineers and geologists.

Decline of Immunity

A significant growth factor has been the gradual stripping away of immunity from the public agencies who design, operate, and maintain our vast system of public works. Recent legislation and court decisions have combined to expose public agencies to a degree of liability unheard of as recently as 20 years ago. Consider these items:

- Under the concept of joint and several liability (commonly referred to as "deep pockets"), it is common for a public agency to have to pay most or all of a decree, even though it may have been judged to be responsible for as little as 1%. This occurs because the other defendants who were judged primarily responsible, such as contractors, engineers, geologists, and developers, may have little money or be grossly underinsured. Some states have recently legislated to some degree against the "deep pockets" concept, but it is still generally prevalent. Even in states where it has been rescinded or modified, the fact that lawsuits can take as long as 5 yr to come to trial means that several more years will go by before the new laws begin to take effect. For obvious reasons trial lawyers who represent plaintiffs are adamantly opposed to modifying "deep pockets." In fact, in California, where most of the state legislators are attorneys, it took a statewide public petition initiative to put such a measure on the ballot, which was then approved by a two-thirds majority vote. Plaintiff trial lawyers argue that their clients will, in many cases, not be able to collect the total true and fair amount of damages if each defendant need not pay more than its proportionate share. This may be true in certain instances. On the other hand, why should a defendant, such as a public agency, be forced to pay more than its proportionate share simply because it has more money or insurance?
- Most plans and specifications for public works and grading improvements, particularly those connected with land development, are prepared by private engineers and geologists hired by developers. The public agency then reviews the plans together with the geologic and

geotechnical reports before issuing a permit for the project. In the past when any of the components of a project failed, liability was borne squarely by the private design professionals and the contractors. Recently there has been an increasing tendency by the courts to hold the public agency at least equally responsible because it approved the documents and issued the permit. As a result public agencies have been fair game for plaintiffs' attorneys. An example of how far public agency responsibility can be extended will be illustrated by one of the case histories on landslides in Chapter 3. Suffice it to say here that a public agency can be held responsible for approving development of a hillside residential subdivision on which a landslide later occurs, even though all the developer's geologists and geotechnical consultants (firms of high professional standing) reported the site to be safe.

Government Mismanagement

Part of the blame for the proliferation of environmentally related lawsuits rests with government itself, from the politicians and administrators to the technical staffs. Unfortunately most government agencies, at all levels, have not fully grasped that they are more vulnerable to liability than they used to be. Some of government's major shortcomings that contribute to or invite lawsuits include:

- Lack of close and competent legal advice in the early stages of planning and design of major projects. Such counsel is essential to seeking out and ameliorating any features of the plans that exhibit inherently high legal risk. The increasingly successful use by plaintiffs of the concept of inverse condemnation (the claim that damage to private property occurred substantially as the result of a government project or action) makes it mandatory to include the study of potential inverse actions as part of the early planning process.

  Example A government agency designed and built a major highway through a hillside area. Where canyons were filled to provide the appropriate gradient and alignment, the agency neglected to design and install subdrains to accommodate the flow of underground water. Concurrently, a large condominium development was built adjacent to and on the uphill side of the highway. The units were completed, sold, and inhabited for several years. Then ground subsidence began to occur within the development, causing major cracking and displacement of walls, slabs, foundations, etc. The homeowners' association employed geotechnical, civil, and geologic consultants to investigate the causes of the settlement and to recommend remedial measures. Many processes contributed to the observed damage, but one cause was the rise of the underground water table due to blockage of the original canyons. If the cause-and-effect relationship can be proven in court, the agency would be liable under inverse condemnation.

This example clearly involves negligence on the part of the agency in overlooking the need for subdrains. However, negligence need not be present to invoke inverse condemnation; it only has to be shown that the damage would not have occurred to the same extent if the agency's project or improvement had not existed. In addition, it is legally irrelevant to argue that the improvements resulted in greater benefits than damages. The only way the agency could protect itself would be to prove that the damage that would have resulted before the project would have been equal to or greater than the actual damage.

Example A government agency approved a large single-family hillside subdivision that included a debris basin at the mouth of a major mountain canyon and a large storm drain from the outlet of the basin to an existing flood control channel. All public improvements, including streets, storm drains, and debris basin facilities, were dedicated to the government agency for operation and maintenance. (Dedication, in this context, means that the appropriate government agency accepted the improvement for ownership and maintenance.) The subdivision included a street with a straight alignment on a 15% gradient, which is the maximum allowable. The street began just below the debris basin and terminated in a T intersection about 1000 ft (305 m) downhill. Opposite the intersection and directly aligned with the end of the steeply inclined street was an expensive two-story house that was not part of the subdivision. A fire in the watershed, followed in the same year by a flood flow with a 25-yr return frequency, caused the basin to become filled with debris. The storm drain became almost completely blocked by a large rock that lodged in its entrance. In this dangerous situation a huge boulder was carried out of the basin by flow diverted over the blocked drain. Having nowhere else to go, the massive rock bounded down the middle of the street, taking dead aim at the house. The ensuing demolition was devastating. Only the foundations were left intact, as the boulder barely slowed down on its course to its final resting point in another debris basin immediately downstream from the house. It was extremely fortunate that no one was home at the time. That the agency owned, operated, and maintained the debris basin, storm drains, and streets established another clear-cut inverse condemnation claim on the part of the homeowners. Studies showed conclusively that, if the subdivision and debris basin/storm drain had not been built, the natural canyon watercourse would have carried the flow, including the boulder, past the house without damage.

• Not learning from a bad experience. In the previous example the government agency, having settled out of court with the plaintiff whose house was demolished by the boulder, chose not to buy or condemn the property and make it part of the lower debris basin. Instead it opted to grant a building permit to the owner to reconstruct the house. All this even though there was no economically feasible solution for averting an identical disaster. Unfortunately the annals of environmental disasters are filled with repeat performances.

- Taking over facilities for operation and maintenance built by other jurisdictions or private parties that do not meet the standards and specifications of the government agency.

  Example A mountain road was privately built to serve a residential development. Several years later a government agency accepted dedication of the road and it became public. At a particular location the roadway had been excavated into a hillside, creating a cut slope that was too steep in terms of the prevailing geologic conditions. No slope failure problems were encountered for many years. In the meantime houses with sewage septic systems were being built upslope. After several years of unusually high rainfall infiltration, combined with sewage effluent from the septic systems, an ancient landslide was activated. The cut slope above the road failed. Certain homeowners sued the agency, alleging, among other things, that the road failure triggered the entire slide and that the slide would not have occurred, or have been as damaging, if the road slope had been up to standard. Regardless of the validity of the allegation, the agency was faced with a costly and time-consuming defense.

  In all fairness it should be said that many government agencies have learned this lesson the hard way and no longer accept substandard improvements until they are properly upgraded. However, there are so many existing substandard dedications that it will be many years before they cease to become targets for lawsuits. In most jurisdictions the present cost of upgrading is prohibitive except when conditions of vulnerability are obvious and extreme.

- Failure to strengthen outdated codes, standards, design criteria, and written policies related to design, construction, and maintenance, *which, when updated, must be rigidly enforced.*

Slosson's law, a corollary to Murphy's law, states that "the quality of professional work will sink to the lowest level that government will accept." Engineers and geologists are therefore often faced with the task of convincing a developer client that quality should be higher than minimum code requirements, or what one can get away with.

Rather than cite examples of such shortcomings it seems appropriate to laud a government entity that has served as a model for how to do things right. The City of Los Angeles encompasses over 1000 mi^2 (2600 km^2), much of it located in environmentally high risk areas. While experiencing unrivaled growth pressures, it has managed to minimize failures due to ground movement and flooding and subsequent lawsuits. The praise should not be construed as approval of all the development that has taken place but rather as acknowledgment that almost all recent development within geologically sensitive and flood-prone areas has been properly engineered, reviewed, constructed, and inspected. This happened because the city revised its old code and adopted and enforced a tough new grading ordinance in 1963. Exceptions were not granted without approval of comprehensive geotechnical and geologic reports. Impartial review boards, composed of eminent practicing engineers and geologists, were established to consider requests for code variances.

The role of political pressure in forcing approval of developments in high-risk areas was drastically reduced. Field inspections and code enforcement have been remarkably consistent, considering that in many jurisdictions these important phases are lax at best and bribery riddled at worst. As the result of these policies the City of Los Angeles has experienced a dramatic 97% decrease in failures due to land movement since 1963. This figure is even more remarkable considering that many of the post-1963 failures were on properties where design and construction had occurred prior to 1963. Two other shortcomings that contribute to or invite lawsuits are:

- Lack of high-quality professional and legal input relative to maintenance and operational programs and procedures.

  Example   A debris basin was privately designed and built to government standards and specifications in conjunction with a residential subdivision. Upon inspection and approval the government agency took over the basin for operation and maintenance. What schedule and policies should be followed in clearing the basin? For instance, what if a storm occurs when the design capacity of the basin is only 50% of maximum (resulting from lack of maintenance following the last storm) and the basin overflows, causing damage to downstream properties? The property owners sustaining the damage then sue the government agency, alleging improper maintenance procedures and claiming, with the aid of their experts, that if the basin would have had 90% of its maximum capacity available, the overflow would not have occurred. It is then incumbent upon the agency to try to prove that its maintenance practices were reasonable and prudent and that not enough time lapsed between the storms to permit effective cleaning. It is obvious that the agency would have been in a far superior legal position if it had formulated, with professional and legal input, a written program or statement of policy in advance of assuming ownership of the debris basin. Even more important, the overflow might have been avoided if not lessened.
- Unwise and environmentally risky projects that result from demands by politicians and that require concessions to sound engineering and geo-

logic practices. Locally elected representatives are particularly guilty of exerting such pressure since land developers usually play a large part in financing their election campaigns.

The Scarce Land Syndrome

One of the paramount reasons for the increase of environmentally related lawsuits stems from the fact that in certain sections of the country, particularly the urban west, much of the prime real estate having few inherent natural risks is already developed. What generally remains is hilly or mountainous land subject to landslide activity, low-lying land subject to flooding and/or subsidence, alluvial fan areas subject to debris and mud flows, and land encumbered by geologic hazards such as seismically active faults and zones of weakness within the rock that may lead to slope failures. Combined with this scarcity of risk-free land is the inflated social status of living in hillside terrain, with its natural beauty, panoramic views, and increased privacy.

In general there has been inadequate engineering and geologic planning of these high-risk areas in terms of flood control, subsidence mitigation, and historic landslide activity. An ideal master plan would identify those areas where hazard mitigation is not economically or environmentally feasible and which should therefore be preserved as open space. This is difficult to do on a project-by-project basis without an overall plan for guidance. The Palisades Highland development in the Pacific Palisades section of Los Angeles is a prime example of such a master plan; it has been preeminently successful in mitigating risks, providing open space, creating an environmentally appealing place to live, and making a lot of money for the developers.

Local government grading ordinances for hillside terrain require that grading and drainage plans by a state-registered civil engineer be accompanied by geologic and geotechnical reports prepared by registered professionals. In essence, the reports are first supposed to identify and evaluate the geologic and soils hazards that could affect the site, such as faults, seismic activity, landslide and mudslide potential, uncompacted fill, soil types, subdrainage, and water-table levels. Second, the reports are to make specific recommendations concerning the proposed grading plan regarding mitigation measures for adverse conditions, treatment of cut-and-fill slopes, buttress fills, removal and recompaction of certain material, subdrain facilities, etc. The agency staff then reviews the plans and the geologic and geotechnical reports for conformance to code, master plan, and other known related reports it has in its files (such as those by the U.S. Geological Survey [USGS]), as well as for general completeness. Review usually results in some corrections and questions, but generally the agency depends on the professional competency and integrity of the state-licensed consultants as to the accuracy of the plans and calculations, thoroughness of the investigations, and wisdom of the final recommendations. Unfortunately the state provides little if any quality control over those it licenses.

Dishonest or Incompetent Civil Engineers and Geologists

At the risk of denigrating our professions and alienating our peers, we feel that the dishonesty and incompetency of some in our fields are the prime reasons for increased environmental-type litigation. Sometimes these failures are closely linked to government mismanagement and scarcity of risk-free land, but more often than not they stand alone.

Example   A land developer purchases a large parcel of unimproved hillside property for the purpose of creating a residential subdivision. The developer immediately employs a geologist to make a preliminary investigation of the site. The geologist reports back that there is evidence of a massive ancient landslide that has not moved for thousands of years but that could become active with the introduction of additional groundwater generated by the proposed development. To determine the degree of hazard and the feasibility of mitigation measures, the geologist recommends employment of a competent geotechnical engineer to assist in extensive soil borings, physical testing, and stability calculations. Knowing that geology is not an exact science and believing the geologist to be overly conservative, the developer consults a second geologist and if necessary a third and a fourth, until a more favorable report is forthcoming. In actuality the time and expense involved in securing multiple opinions are not usually necessary. Most developers and their attorneys are aware of those geologists who are most likely to generate an optimistic opinion.

Once the final report is submitted, the public agency has little option but to approve it. There may be some questions and doubts, but in most cases the developer is ultimately given a green light. First, the agency usually has neither the budget nor the political authorization to perform an independent physical investigation. Second, and more important, the agency is caught in a dilemma. If it does not issue the permit, the developer will probably sue. After all, the only report on record is that of a state-licensed (and therefore presumably qualified) geologist, and denying a permit would be tantamount to condemning the land without compensation.

On the other hand, issuing the permit will avoid legal consequences now. But what happens 10 yr from now if a damaging landslide does occur? The developer may be out of business and the geologist may have no assets in his or her own name and very little, if any, errors-and-omissions insurance. Besides, the geologist would probably only have to argue that the government accepted the report and that the recommendations, therefore, met standard practice.

So who becomes the "deep pockets" victim stuck with paying the major share of a successful court judgment? The taxpayers who support the agency, of course. What makes it such a bitter pill for the taxpayers to swallow is that a typical verdict judgment may exceed by several hundred times the cost of defending an initial lawsuit or actually buying the unimproved property. And this does not take into account insurance company losses, which are reflected in high premiums, or, most important, potential injury and loss of life in the landslide.

If the foregoing scenario sounds a bit contrived and melodramatic, consider this: The authors have been directly associated with approximately 15 lawsuits won by plaintiffs against government and totaling about half a billion dollars in monetary judgments or settlements. In every single one of these cases there had been geologic and geotechnical reports written by licensed professionals that, in effect, declared the site either nonhazardous or capable of being rendered safe by relatively simple mitigative actions.

Lest the reader gain the impression that we are making a blanket indictment of all consulting civil engineers and geologists for opening the door to litigation by their performances, we should point out that the guilty ones are relatively few. But, in any given metropolitan area, the same names keep popping up again and again, and these individuals are seemingly immune to personal liability and oblivious as to how their conduct is adversely affecting their professions and the general public. Some of these individuals and firms have as many as 20–30 pending legal actions against them at any one time.

All of the preceding factors have influenced the burgeoning growth of environmental forensics. The growth represents an unhealthy trend for the civil engineering and geology professions, as well as for society as a whole. It is up to professionals to do their part to slow the trend by doing quality work and by not yielding to the temptations provided by their clients.

The authors have come to believe that most environmentally related accidents, disasters, and failures are not "acts of God." If one conducts a thorough and competent forensic investigation, the underlying causes are nearly always human rather than natural, and in almost every instance the event was recognizable, predictable, and preventable if only good professional judgment had been used.

The Role of the Expert Witness

The public, as a whole, has a huge financial stake in the outcome of lawsuits involving environmental disasters and failures. It is in society's best interest that justice be served in the courtroom. Too often it is not. There is a better chance of justice prevailing if both sides are represented by honest, well-qualified experts who present a thorough and objective picture of all the pertinent factors. Many bad laws and unfortunate legal precedents have resulted from erroneous, devious, or incomplete testimony from experts.

An unjust verdict against a public agency, public utility, or insurance carrier must ultimately be paid by the general public in the form of increased taxes, decreased services, or increased utility and insurance rates. It is our observation that juries, who once emotionally sympathized with the "underdog" plaintiff against the "big bad" public and corporate entities, have recognized this fact. Consequently, we believe juries are becoming much more objective today.

In the case of a just or even partly just verdict against a public agency, the taxpayers can, through their elected representatives, pressure the agency to take the necessary steps to insure that the disaster (or similar disasters) does not recur.

There is the possibility, or at least the hope, that politicians, public administrators, and professionals who are faced with similar potential problems in other geographic areas can learn and benefit from a trial that has good technical input, regardless of the outcome.

Probably the most compelling argument for the need of expert witnesses is to imagine what it would be like without them. Does anyone really want the fate of precedent-setting environmental cases to be determined only by lawyers or politicians?

Forensic Work Is Not for Everyone

Having great experience and eminent qualifications in a particular field does not necessarily imply that one would be an effective forensic expert. The investigative phase of the work would probably be easy for such a person, but the litigation aspect could prove difficult. There are several traits that characterize good expert witnesses:

1. The ability to explain the most complex technical concepts in terms easily understood by judge and jury. Often brilliant academic or research scientists are unable to explain their findings and concepts in elementary terms, as one would lecture to a freshman college science class composed of general education students.
2. The patience to educate their client's attorneys in all the technical nuances of the case.
3. The temperament to remain cool and objective when under attack by opposing attorneys.

In time, the first two traits can be developed and honed. The third comes more naturally. A short fuse can be exploited to great advantage by a clever lawyer. Many engineers and geologists find that they do not like forensic work for this reason.

Another aspect of the expert witness' role that can be uncomfortable is the prospect of testifying against another civil engineer or geologist. Merely being opposing witnesses does not usually present a problem because of the controlled adversarial environment in which these confrontations take place. However, to testify as to the accuracy or judgment involved in the actual work product of a peer is another matter. The prospect of making a public judgment regarding a peer's work can be a little frightening and can cause some deep soul searching. Even in the event of a mistake so flagrant that it is not subject to controversy, there persists the thought, there but for the grace of God go I. In many instances the error in mechanics or judgment is not clear-cut. Often it all boils down to a rather vague criterion, such as whether it was "good engineering or geologic practice"—a difficult yardstick against which to impugn a person's professional reputation.

In short, being an expert witness is serious business and one should weigh these considerations carefully before embarking on a forensic assignment.

Qualifications

Most civil engineers and geologists gravitate to forensic work late in their careers. Some of the primary reasons are

1. They are attractive to attorneys because of their great experience and high stature in their professions.

2. Most have specialized in certain fields. Specialists are more in demand than generalists. (There are a number of enterprising organizations that act as brokers, providing attorneys with names of their members whose technical backgrounds match the specialty requirements of a particular lawsuit.)

3. Forensic work can be ideal for someone who has retired but wants to continue to work part time. Often it requires little or no staff and minimum office overhead.

4. It can be both interesting and challenging and is usually a radical departure from a routine that may have become stale—almost like a second career. In many instances the research and discovery involved make it similar to a graduate-school research project, where experience and knowledge can be combined to produce new or improved technology that, in turn, may help to improve the standard of practice.

5. Unlike design and construction endeavors, forensic work is *relatively* free of liability. After all, it is hard to imagine being sued for rendering an opinion. However, an expert witness is not completely free from liability exposure, as illustrated by a personal experience. One of the authors had acted as an expert witness for a defendant in a major lawsuit. The defendant won. Meanwhile the attorney for the plaintiff was so upset with the performance of his own expert witness that he refused to pay the full amount of the

expert's billing. A lawsuit was then initiated by the expert against the lawyer to collect his remaining fees. This action so infuriated the attorney that he called the author to ask if he would serve as an expert witness in a countersuit charging incompetence. When the author declined the request the lawyer intimated that he could subpoena the author, thereby avoiding the payment of expert witness fees. This was not necessarily true because, under subpoena, testimony would have been confined to an eyewitness rather than an expert witness nature. Fortunately the author was not forced to testify. The moral of the story is that it is possible to be sued for being a poor expert witness under extreme circumstances, but highly unlikely. Therefore one of the advantageous features of forensic work is that it probably does not require expensive errors-and-omissions insurance.

6. Last but not least, forensic work usually pays well. Forensic experts work by the hour, typically one rate for deposition and trial testimony time and a lower rate for investigative and consulting time. Any type of incentive fee arrangement, in other words, any payment based on winning the case or on the amount of the award, not only would be unprofessional but, if discovered by the opposition or the court, would completely discredit the witness.

Plaintiff or Defendant?

In general, is it better to be an expert witness for a plaintiff or a defendant? Many experts do seem to gravitate toward one side or the other. The reason may be linked to the fact that many attorneys represent either plaintiffs or defendants almost exclusively.

Once an expert is employed successfully by a public agency, insurance company, or large law firm, word of such success spreads to other attorneys within the same or similar organizations. Also, most experts are naturally reluctant to testify against someone who has become a regular client, no matter what the merits of the particular case might be.

A broad generalization (with numerous exceptions) can be made that it is more difficult to be an expert for a plaintiff than for a defendant. The reasons revolve around the fact that most plaintiff attorneys are compensated on a contingency basis, whereas most defense attorneys are on salary or paid by the hour. An attorney working on contingency has a huge financial stake in the outcome of the trial, as well as having his or her professional reputation at stake. Of course an attorney working on salary or by the hour also wants to win; winning is the name of the game and a basis for advancement in trial law. However, there is not the same direct financial pressure to win. Consequently, in representing the plaintiff, an expert is more likely to be working with tighter budgetary constraints on investigations. There might also be more pressure, albeit subtle, on the expert to compromise his or her testimony to fit a prepared script. Working for the defendant, on the other hand, usually allows for more time and a greater budget for investigation. Furthermore, most defense attorneys are anxious to know the facts so that a settlement can be reached if the defendant's case is not strong.

In our opinion, the best approach to this question is to accept an invitation to an assignment based on a preliminary investigation of the facts of the case and not on whether it is with the plaintiff or defendant. A reputation for having worked for both sides is an indication of fairness and objectivity to the judge or jury. Most previous clients recognize this and are not rancorous or spiteful to see an expert working for the other side. The authors have spent approximately one-third of their forensic careers working for plaintiffs and two-thirds for defendants.

Agreement with Client

Before accepting a forensic assignment it is wise to complete a routine checklist, particularly with an unknown lawyer or law firm.

1. Be certain there are no conflicts of interest with any of the parties connected with the lawsuit, directly or indirectly.
2. Obtain as much background information on the disaster or failure as possible, mindful that such information may be biased or lack pertinent data.
3. Attempt to separate facts from opinions so as to be able to form an objective early picture of the main issues.
4. Inquire as to the status of the case and its tentative schedule to determine whether there is sufficient time for a thorough investigation.
5. Make certain the subject matter is within the appropriate areas of expertise for you or your firm. The attorney should clearly understand which issues are within that province and which are not.
6. Discuss fee schedules and determine when and by whom payments are to be made.
7. Check out the reputation of the attorney or law firm if it is not already known.

A written and signed agreement with the client is always good practice. It need not be elaborate but should cover the scope of the work to be performed, time schedules, and fee arrangements. In most cases it is difficult to estimate the total fees in advance. However, estimates that are subject to periodic revisions are a good idea. They keep the client up to date on projected expenses and the consultant on track as to the scope of work revisions or additions.

2 INTRODUCTION TO CASE HISTORIES

The case histories that follow were selected with three objectives: first, to present as many diverse types of flooding and earth movement forensic investigative experiences as possible; second, to choose cases that would exhibit the maximum number of technical, legal, professional, and ethical issues that an expert might encounter; and third, to select cases that would spark the greatest reader interest relative to their scope, universal applicability, and importance in establishing precedent.

All of the cases are rooted in California except one, which is also set in the west, in Utah. However, if the old axiom about California possessing the dubious distinction of being the nation's trendsetter applies to forensic engineering and geology, there is no reason to believe that similar types of cases have not or will not crop up elsewhere. The technical and legal problems engendered by floods and earth movement are certainly global in scope.

We have reconstructed the case histories from our own files, records, and memories. The reader should bear in mind that they are written from our personal viewpoints and that the editorial content is entirely our own. Actual names of people and organizations are not included, with the exception of public agencies and in certain instances where we felt names were appropriate.

Each of the case histories is separated into four basic components:

1. "The Event," which is intended to provide a factual account of the disaster or failure
2. "The Investigation," which describes the investigative routine we and our colleagues followed, our analysis, and our conclusions
3. "The Litigation," which is our personal view of the legal proceedings from the time we were employed through trial or settlement
4. "The Post Mortem," essay in nature, in which we address such subjects as how the disaster could have been avoided or mitigated, how future similar disasters might be averted, and, in a broad sense, what might be learned from the experience

Chapters 3, 4, and 5 address the subject of landslides and related earth movement; failures that result in annual nationwide losses of nearly $2 billion. The three case histories include:

1. A landslide in a posh, 280-home residential area that resulted in considerable damage and gargantuan lawsuits. Whereas some of the homes were completely destroyed, most sustained a level of damage that, after dewatering and stabilization, could easily be repaired. The landslide has been at least temporarily arrested by maintaining a lower water table through the use of wells. A comprehensive program to attempt to achieve long-term stability is still being formulated. The case history is rife with important principles— geologic, political, and legal—and serves as a classic example of how to invite lawsuits.

2. A landslide in a remote area having no structures atop the slide mass. However, the damage that ensued below from slide debris was of enormous proportions. A major railroad and interstate highway were engulfed by the debris, which formed a natural dam, backing up river flow and inundating a small town. It was necessary, at tremendous expense, to construct tunnels through a mountain to reroute both the railroad and the river. Many landslides are caused by human activity such as improper grading and introduction of additional water. This one, as in most, was triggered by natural precipitation falling on a quasi-stable land mass. The basic legal issue was whether the railroad geologists should have recognized the potential problem and recommended relatively inexpensive mitigation measures to protect the railroad.

3. A landslide that, like the first, occurred in an affluent hillside residential area and was caused by similar groundwater-related problems. As with the first slide, the lawyers sprang into action while the slide was still in motion. However, the difference in this case history is that the litigation was suspended while a panel of experts worked out a permanent mitigation program to be paid for by the defendant, a public agency. If successful, not only the homeowners would be protected, but the taxpayers would save millions of dollars.

Chapter 6 is the forensic study of a damaging flood within a natural floodplain (used primarily for agriculture) that had exhibited a long history of flooding. What made this particular flood so controversial and litigious is that it:

- Occurred after completion of certain upstream flood control improvements
- Was international in scope and involved actions not directly controllable by the entities who were damaged
- Was partly the result of land-use planning decisions that may have been made with insufficient consideration of the legal consequences

The fundamental forensic question posed in the case was whether certain properties were damaged more severely by the flood than they would have been had no upstream flood control facilities been built. It proved to be an investigation fraught with technical complexities and international complications, and, in the search for answers, it uncovered societal issues much broader in nature than the case itself.

The case history contained in Chapter 7 is an account of the sudden inundation of a residential neighborhood that could not be explained on the basis of usual hydrologic and hydraulic criteria and calculations. Nearly all the inundation occurred in a nonfloodplain area and was caused by an odd succession of events and circumstances. After all the forensic data were analyzed, the preponderant legal question was whether the combination of events constituted an unavoidable act of God or whether, despite its bizarre nature, the disaster could have been avoided with more prudent decisions and actions by the government entities involved.

No lives were lost in the events in the previously described case histories. However, the fire–flood sequence case history in Chapter 8 was a profound human disaster involving the abrupt and violent deaths of 13 people whose bodies were swept away and never recovered. The principal factors contributing to the demise of the tiny mountain resort community included a large watershed, steep terrain, highly erodible soil and rock types, and a wildfire that consumed nearly the entire watershed, capped by a storm the following winter of a 25-yr magnitude return frequency. Erosion at the height of the storm was so severe that the peak discharge in the main canyon contained about 20% water and 80% mud, rock, and debris. The legal controversy centered on whether any of the government agencies having jurisdiction in the area could, by any reasonable action, have either mitigated the effects of the flash flood or given earlier warning to the residents.

The gradual subsidence or areal settlement of large land masses is as old as recorded history. Chapter 9 deals with the consequences of such settlement in a heavily populated coastal area. Settlement essentially occurred during a 20-yr period when a much greater volume of groundwater was being extracted by wells than was being replaced by natural processes. The magnitude of that settlement reached as much as 6 ft (2 m). Because the movement was so gradual it did not cause a great deal of damage to either structures or underground utilities. However, the section of the city adjacent to the ocean bay was profoundly affected. The area, which had historically been several feet above mean sea level, had sunk to several feet below, making it extremely vulnerable to flooding and tidal action. A combination of flood flow and abnormally high tides caused the inundation and led to the ensuing litigation, which is the subject of the chapter. The case history deals not only with the consequences of excessive groundwater extraction but with public policies of flood control priorities, reservoir-level management, and other related issues.

River morphology is the key issue in the case history described in Chapter 10. Valuable fruit-orchard land was being lost by riparian property owners as a result of collapse of their riverbanks. In seeking the causes of the bank failures, the forensic investigators focused on sand and gravel mining activities taking place in the riverbed 6–8 mi (10–13 km) downstream. Our studies indicated that the lowering of the riverbed due to gravel extraction was not being fully replenished by natural bed-load transport and deposit. Because the lowering of the streambed at the gravel pits moved inexorably upstream, the deepened channel caused the side slopes to become steeper than their natural angle of repose and, hence, collapse. Experts representing the gravel operators argued that the deeper channel adjacent to the orchards was entirely the result of natural processes and was not related to the gravel mining.

3 MALIBU LANDSLIDE: MASSIVE LITIGATION

The term *landslide* is defined by the USGS as "a wide variety of processes that result in the downward and outward movement of slope-forming materials composed of rocks, soils, artificial fill, or a combination of these. The materials may move by falling, toppling, sliding, spreading or flowing." This broad description covers all the types of movements shown in Table I.

The landslide that is the subject of this chapter and the landslides described in Chapters 4 and 5 are identified as "slides" in Table I. They were all characterized by mass movements in which there was a distinct surface (or surfaces) of rupture or a zone of weakness that separated the slide material from the more stable underlying material.

THE EVENT

The long and tortured history of this 1983 Malibu, California, landslide began for one of the authors in 1962 when he was a neophyte geology consultant. At that time the owner/developer of several hundred largely unimproved hillside building sites, located on what is known as Big Rock Mesa (Fig. 3–1), employed the author to perform an engineering geologic investigation and report. The report was required for issuance of a permit for the grading and development of a residential subdivision of the property.

In the course of his geologic field investigation, the author discovered what he considered to be evidence of a deep-seated ancient landslide underlying Big Rock Mesa and the property, a feature not then shown on the usually reliable USGS maps. Still acutely naive concerning the ways and wiles of big-city developers, he disclosed his suspicions to the developer and the other consultants at a project meeting. To say that all hell broke loose would be a gross understatement. The author was instantly fired and told in no uncertain terms to destroy all written notations pertaining to his wild, unfounded theory. Needless to say he was not paid for his services and was threatened with a lawsuit if there were any leaks to the county or media. In fact, he was so embarrassed by the experience and tormented by the thought that he might have been wrong that he never pressed for collection of his fees.

TABLE I
Abbreviated Version of Classification of Slope Movements (Varnes, 1978)

TYPE OF MOVEMENT			TYPE OF MATERIAL		
			BEDROCK	ENGINEERING SOILS	
				Predominantly coarse	Predominantly fine
FALLS			Rock fall	Debris fall	Earth fall
TOPPLES			Rock topple	Debris topple	Earth topple
SLIDES	ROTATIONAL	FEW UNITS	Rock slump	Debris slump	Earth slump
	TRANSLATIONAL	MANY UNITS	Rock block slide / Rock slide	Debris block slide / Debris slide	Earth block slide / Earth slide
LATERAL SPREADS			Rock spread	Debris spread	Earth spread
FLOWS			Rock flow (deep creep)	Debris flow (soil creep)	Earth flow
COMPLEX			Combination of two or more principal types of movement		

Twenty-one years later the author was resoundingly vindicated. However, there could be little personal satisfaction derived from a catastrophe that eventually resulted in hundreds of millions of dollars in property damage, inflicted incalculable emotional trauma on hundreds of homeowners, and spun a vast, complex, and costly web of lawsuits and countersuits.

Famous for its large colony of film stars and other show business personalities, Malibu also has a well-deserved reputation as the natural disaster capital of Southern California owing to its unique combination of geography, topography, geology, and weather characteristics. The community of Malibu consists of an elongated strip of expensive real estate, hugging 6 mi (9.7 km) of the Pacific coastline between the City of Santa Monica and the Ventura County boundary. In a 1990 special election Malibu voted to incorporate as a separate city, thereby divorcing itself from Los Angeles County.

The perpetually gridlocked four-lane Pacific Coast Highway (PCH) splits Malibu topographically. Seaward of the highway lies a narrow, heavily populated coastal zone where the topography is generally flat and little higher than sea level. This area is typically so narrow that there is space for only a linear row of lots fronting the ocean and backing, with garages, the PCH. Immediately inland from the highway the land rises abruptly in steep cliffs topped by mesas and rugged mountain terrain. Some of the mesas, such as the one known as Big Rock, have been densely developed.

Even some of the steep mountainous areas have been graded to accommodate development. Pepperdine University, considered a jewel of physical

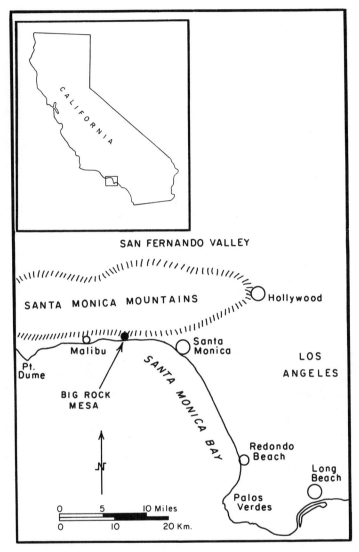

FIGURE 3–1
Vicinity map of the Big Rock Mesa landslide.

beauty among college campuses, was carved from such terrain. The property seaward of the PCH is subject to severe battering and flooding by the surf anytime nature produces a combination of heavy storms, crashing waves, and high tide, a common phenomenon well publicized by resident Johnny Carson.

A 1962 oblique aerial photograph of Big Rock Mesa before development. At that time it was the opinion of one of the authors that the large amphitheater-shaped mesa in the center of the photo displayed the morphology of a subdued or ancient landslide.

In addition, some of these same oceanfront properties can be damaged from the highway side by sloughing of the bluffs due to erosion, rock fall, and landslide activity. Closing or partial closing of the highway due to slope failure has become a way of life for Malibu residents.

From a geologic standpoint the land lying inland of the PCH is generally as unstable in terms of development as one can find anywhere. Foremost among the many problems are the overly steep bluffs, highly fractured rock, adverse bedding planes, active and inactive landslides, and active and inactive faults. All these potential geologic factors contribute to an environment ripe for damage from earthquakes, landslides, debris flows, and excessive erosion.

In addition to the inherent problems of land and sea, Malibu is confronted

Oblique aerial view from the west showing the Pacific Coast Highway, the height and steepness of the bluff (caused by wave erosion), and the narrow beach area.

with a constant fire peril owing to the brush-covered mountain slopes, which become tinder dry by midsummer. When the fires are followed by high-intensity storms, a dangerous fire–flood sequence can occur, as described in Chapter 8.

Given the wide assortment of latent disasters it seems strange that anyone would want to live in Malibu. Just the opposite. Residents are a hardy lot who appear willing to trade the potential dangers for the prestige of residing in a celebrity-studded community that offers either great beachfront living or spectacular ocean views. In addition to being hardy (some might say fool-hardy), Malibu residents are fiercely independent and more than a bit stubborn. Despite the long-recognized health and land-stability problems posed by the use of individual septic tanks and seepage pits, the residents have doggedly refused to permit construction of a sanitary sewer system to serve all of Malibu. Three times in the past 25 years the county has put a sewer bond issue on the ballot, and three times the residents have voted it down. Even today, as Malibu achieves cityhood, the sewer issue persists.

There is no doubt about the reasons for the long-standing hostility toward sewers. The huge cost, which would be assessed to all benefiting property owners, has certainly been a factor. But far more compelling is the mere desire to keep Malibu rustic and rural with little or no future growth. Sewers, it is feared, would encourage additional growth, and well they might. To some extent residents cannot be blamed for having adopted this stance. New growth would place a heavy burden on an already overloaded PCH as well as on the region's fragile environment. We emphasize this sewer issue because, as described later in this chapter, individual septic systems have played a major role in raising groundwater in certain Malibu areas, to the extent that they have been a major factor in causing landslides, including the one at Big Rock Mesa.

Big Rock Mesa was ripe for development in the early 1960s. Located adjacent to the PCH for easy access, it possessed an uneven, sloping mesa about 300 ft (90 m) above sea level that could be graded with relatively modest volumes of earthwork to provide individual homesites, almost all with panoramic ocean views.

The geotechnical engineering and geology professionals who were employed in the design of the subdivisions in 1963 seemed aware of the more obvious dangers posed by the marginal stability of the bluff, but apparently did not recognize, or chose not to recognize, the potential for a deep-seated landslide. Los Angeles County officials approved permits for the subdivisions, which included individual septic systems, based on assurances of a 1.5 factor of safety for bluff stability. One of the conditions of approval, as recommended by these consultants, was the installation of 2800 ft (840 m) of horizontal drains (hydraugers) to help dewater the mesa. Hydraugers were to be operated and maintained in perpetuity by an association of the homeowners, but no regulatory authority was assigned to the county to assure compliance.

Development began in 1964, and by 1983, 280 homes had been built on the mesa. Meanwhile, in 1975 the homeowners acted to dissolve the association, and the hydraugers, as well as several pumping wells, were abandoned.

As early as 1978, a year of record high rainfall, some of the residents reported evidence of land movement. Despite these early warning signs, favorable geologic reports were still being submitted by consultants and approved by the county. Development of the mesa continued unabated. In 1980 there were more abnormally high volumes of rainfall, causing groundwater levels to rise still higher and producing further evidence of ground movement in the form of cracks and ruptures. Still, one more home—in the $2 million category—was built. Ironically, this was one of the first homes completely destroyed in 1983, and its owner was the sole plaintiff in the initial lawsuit.

In September 1983, following another season of record rainfall, the increase in the volume of groundwater and development of pore water pressure was enough to tilt the driving forces (combined with the ever-present force of gravity) over the forces of resistance, and the whole mesa began to move seaward. The surface area of the slide block encompassed about 150 acres

A 1983 oblique aerial view of Big Rock Mesa with the Pacific Ocean to the left (south). Note the bluff has a scalloped appearance from numerous small landslides and slumps. The back scarp of the main landslide is located upslope of the road in the center of the photograph. The three parts of the Big Rock Mesa landslide can be recognized by the slight lows in elevation and the shadow pattern. Photo courtesy of Geo-Tech Imagery, Oceanside, California.

(60 ha) and involved about 216 homes. Depth of the slide was yet to be determined, but the limited available subsurface data suggested a depth as great as 200 ft (60 m). Movement was of a very gradual nature, with the slide creeping inexorably seaward at a rate of approximately 0.125 in. (0.32 cm) per day. Within several months at least six homes were so badly damaged that they had to be abandoned. Most of the others sustained some degree of damage but were eventually able to be restored. As the toe of the slide moved into the surf zone, homes along the seaward side of the PCH were uplifted several inches. The initial estimate of residential damage was about $200 million.

Water mains were severely damaged. Repairs and replacement by the waterworks district cost several million dollars. In order to gain access to water-line ruptures and leaks, some of the larger mains were reinstalled above ground rather than buried in trenches. Streets and storm drainage facilities also required constant inspection and repairs during this period.

There was no doubt among all the geotechnical engineers and geologists, both government and private, that the only way to attempt to arrest the slide was to lower the groundwater table as quickly and as drastically as possible. To this end the Los Angeles County Board of Supervisors created a $1.3 million County Improvement District in December of 1983. The district's groundwater extraction program consisted of installing 18 groundwater pumping wells, cleaning and reactivating the abandoned hydraugers, and adding hydraugers at strategic locations. When completed, peak production of the dewatering system approached 450,000 gal (1,705,000 liter) per day, which was sufficient to lower the water table substantially and to slow the slide movement.

By early 1984 those residents whose homes had not been severely damaged breathed a collective sigh of relief. It had been a terrible ordeal. Some mental health professionals compared the emotional trauma suffered by Big Rock Mesa landslide victims with that of battle fatigue during a war. Unlike quick-hitting disasters, this slide subjected the victims to a prolonged period of intense emotional stress as they helplessly watched their homes slowly disintegrate. There was also the ever-present fear that the slide would suddenly accelerate, becoming life-threatening before they could evacuate. A Malibu psychiatrist, himself a victim, eloquently described in a newspaper interview the agony of trying to sleep while listening to his house groan, creak, and crack.

After the immediate danger had been minimized there was still a long-range mitigative effort to face.

THE INVESTIGATION

In early 1985 the authors were invited by a law firm representing one of the three defendants, Los Angeles County Waterworks District No. 29, to perform forensic civil engineering and geologic investigations in connection with the landslide. A geotechnical engineer and groundwater hydrologist also were retained by the district and added to the forensic team. The other two defendants were Los Angeles County, including the Los Angeles County Flood Control District, and the State of California Department of Transportation (CalTrans). Each defendant employed its own team of experts because of the possibility that conflicting interests might develop. We learned that the strategy of the homeowners' attorneys was to file a lawsuit on behalf of only one of the homeowners. It was to be, in effect, a test case and thus extremely important for both sides.

We undertook our assignments with the understanding that our investigations would be scientifically objective and would encompass all facets of the causes and effects of the damage to the plaintiffs' properties, not just those related to the role of the waterworks district. The importance of a complete investigation cannot be overemphasized. Often an attorney will instruct an expert to confine the investigation, and therefore the testimony, to certain carefully selected issues. There are usually two reasons for such a request: to save money and to hide or obscure facts that might prove damaging to the attorney's case.

An example of such obfuscation occured in a case in which we were personally involved where a farmer, whose property was located downstream in an agricultural floodplain, sued an upstream farmer for grading that allegedly increased the volume of sediment deposited on his property during a high-intensity storm and flood. The plaintiff's expert civil engineer, a person of high professional standing, was asked by the plaintiff's attorney to investigate and estimate the cost of removing the sediment to restore the site to its preflood elevations. The expense of restoration proved greater than the preflood market value of the property. During his deposition the expert was asked by the defendant's counsel if he had estimated the cost of merely releveling the land and resetting the irrigation system, the method by which most farmers in the floodplain have coped with excessive deposition. The expert's answer was no. When asked why not he explained that he had not been instructed to do so. Thus, instead of investigating all methods and costs of mitigation, he had narrowly confined his investigation to a method dictated by the attorney for his own self-serving purpose.

This example is more obvious than most instances of attorneys' instructing witnesses as to the scope of their investigations. We do not mean to imply that there be no definition of the expert's scope of activities. Certainly no work should be done that is irrelevant to the cause and effect of the disaster. In fact, the attorney and expert should have a clear understanding (preferably in writing) regarding the scope of the investigation, and the expert should not deviate from this scope without authorization.

The principal allegation against the waterworks district was that leakage from its water mains in the years prior to September 1983 had infiltrated the slide mass, thereby contributing a significant causation factor to the landslide. It was evident from the outset that there would be no dispute about the basic triggering mechanism of the slide—too much groundwater.

Knowing that water infiltration of the slide mass from all sources was the primary cause, the scope of the civil engineering investigation seemed clearcut. First, it would be necessary to conduct a comprehensive study of the waterworks district's maintenance and repair records to determine when and where leaks occurred, size of the lines involved, cause of the leaks, flow volumes, approximate direction of the leaks, depth of trench, method of bedding and backfill, and any other information pertinent to estimating the total volume of water-main leakage that would have infiltrated the subsoil. The second step would involve estimating infiltration volumes, for the same period of time covered by the leak records, from all other sources; namely, rainfall, septic systems, and irrigation. Whether the infiltration from leaks constituted a significant contribution, in comparison to the other sources, would be a matter for the court to decide. In the process of estimating the volume and timing of all sources of infiltration, useful data would be generated for the other members of the waterworks forensic team.

A tentative format for the geologic investigations was, of necessity, much broader in scope. It included

1. Field reconnaissance of Big Rock Mesa utilizing visual advantage locations

2. Aerial overflight to obtain a clear overview of geologic setting

3. Development of a first data set and the accompanying multiple hypotheses; the data set to include

    (a) Data for interpretation of geomorphology (i.e., aerial photographs)

    (b) Basic geology related to rock type, condition of rock, structure, origin, age, and tectonic effects on the rock materials

    (c) Evidence of past landslide, coastal erosion, and/or tectonic origin of large Big Rock Mesa amphitheater area, with the accompanying coastal bluffs along seaward side. Evidence of reactivation of past landslide, including mapping of boundaries, surface expression, and gathering of kinematic data.

    (d) Beginning of data recordation on geologic maps and cross sections

4. Initial major data search and pertinent data collection and cataloging, including data from USGS, California Department of Mines and Geology (CDMG), Los Angeles County Engineering Geology section, and local universities, with emphasis on theses and dissertations (CDMG's listing of all theses for topics in California to assist in such data searches)

5. Compilation and testing of all working hypotheses

6. Development of a kinematic map depicting the direction of motion and amount of motion for recognizable topographic and physical features (structures, survey control points, prominent trees, etc.)

7. Correlation of failure(s) to climatic changes (i.e., sea-level fluctuations, rainfall)

8. Relative age dating of ancient slide(s) by various methods. Applying/determining probable first age of movement based on geomorphic and climatologic data.

9. Borings and testing in conjunction with the geotechnical engineering consultant to determine depth of slide plane (or planes), direction of movement, etc.

10. Detailed geologic investigation of the plaintiffs' property and environs, possibly involving testing

11. Final tests of working hypotheses to determine most probable cause of landslide(s)

12. Study of potential mitigation measures and procedures

Essentially the geotechnical engineers would do the physical testing of material removed from the borings and perform all strength and stability calculations. Emphasis would center on the relationship between aquifer levels and the factor of safety for stability. The groundwater hydrologist would attempt to establish a water budget between groundwater inflow and outflow. We would coordinate our work to take advantage of each other's expertise and to eliminate noncontroversial duplication where our efforts overlapped. Although we would generally need to agree on data and hypotheses, our conclusions could be somewhat different.

Water-Line Leaks

We found that, prior to development, Big Rock Mesa had been used as a site for farming and cattle grazing. Water for these uses was supplied by private on-site wells. Through the years the wells apparently pumped sufficient water to counteract rainfall infiltration, thereby maintaining a reasonably low groundwater level.

In 1962, coincidental with the first subdivision on the mesa, Los Angeles County Waterworks District No. 29 designed and constructed a public water system to serve the mesa. The system was designed with minimum-diameter (4 in.; 10 cm) mains to provide 1000 gal/min (63 liter/sec) fire flow protection as well as domestic water service.

In reassembling waterworks district files for records of leaks, we confined our search to a 17-yr period from 1967 through 1983; 1967 was chosen as a starting point because development of the initial subdivisions had been completed. We could then compare water-line leaks with all other sources of infiltration for the same time frame.

The search for leak data proved tedious. Several types of documents had been used by the district to record the discovery of leaks and repairs. They included both chronologic telephone logs and repair records, but neither was segregated by geographic area. Thus it was necessary to cull only Big Rock Mesa addresses.

Our research resulted in discovering the following numbers of leaks for the 17-yr period:

Main lines	23
Service lines	64
(main lines to meters)	
Meters	126

Leaks on private property were ignored in the analysis for two reasons: first, because the responsibility for their repairs rested with the individual property owners; and second, because most such leaks were not reported to the district and therefore were nearly impossible to track down.

The records contained very little information as to either the magnitude or the duration of the leaks. Most were probably so minor that they were not quantifiable. However, to be extremely conservative (i.e., to err on the side of making leak volumes appear larger than they really were) we assumed the following:

	Average leak rate (gal/min; liter/sec)	Average duration (hr)
Main lines	200; 12.6	6
Service lines	20; 1.26	4
Meters	5; 0.32	4

Because of the relatively shallow trenches (3–4 ft; 0.9–1.2 m deep), coupled with less than optimum compaction of the trench backfill (hydrocompaction rather than mechanical), it was agreed by our team that an overwhelming proportion of the leakage would rise to the surface and evaporate or run off in the streets. Again, to be conservative, we assumed that 50% of the leakage volume from the pipes and 10% from the meters would infiltrate the subsoil.

Computation of Leak Infiltration

Mains 23 × 200 gal/min × 6 hr × 6 min × 0.5	=	828,000 gal
(23 × 2.6 liter/sec × 6 hr × 60 min × 60 sec × 0.5)	=	(3,130,000 liter)
Service lines 64 × 20 gal/min × 4 hr × 60 × 0.5	=	154,000 gal
(64 × 1.26 liter/sec × 4 hr × 60 min × 60 sec × 0.5)	=	(580,000 liter)
Meters 126 × 5 gal/min × 4 hr × 60 × 0.1	=	15,000 gal
(126 × 0.32 liter/sec × 4 hr × 60 min × 60 sec × 0.1	=	(58,000 liter)
		997,000 gal (3,768,000 liter)
	or	3.07 acre ft (0.38 ha m)

We were to learn later that the hydraulics expert for the plaintiffs did not attempt to estimate leakage volume on the basis of repair records. Instead he drew on his personal experience with other water systems to opine that:

1. Comparable systems in similar terrains experienced many more leaks per mile of main during a given time period than our analysis indicated.
2. The "unaccounted for" water in the Big Rock Mesa system (the difference between the volume registered at the main meter and the sum of registered volume at the users' meters) constituted a reasonable means of measuring leak volume.

Prior to trial a study was conducted by an independent testing company to determine the unaccounted for water in the Big Rock Mesa system. The study concluded that there was an approximate 9% unaccounted for factor, which is not an insignificant number. However, leaks in the system represented only one source of actual or apparent water loss. Other actual losses included hydrant use for fire fighting and unauthorized taps for construction and other purposes. However, the predominant factor in unaccounted for losses was inaccurately registering house meters. In this regard research of several

American Water Works Association technical papers revealed some salient points:

1. Domestic water meters tend to underregister flow volume. The degree of underregistration generally depends on the quality and age of the meter, the quality and frequency of maintenance, and the flow rates. As flow rates increase, the meters underregister less. For instance, at flow rates of 10 gal/min (0.63 liter/sec), meters 8 or 10 yr old generally register about 98% of actual flow. At the low end of the spectrum, flows of 0.25 gal/min (0.016 liter/sec) have been found to register as little as 70% of actual flow. At less than 0.125 gal/min (0.016 liter/sec), some do not register at all. Human nature is involved too. If a customer suspects an underregistering meter as the reason for a suddenly decreased water bill, chances are slim it will be reported. The opposite suspicion, of course, would ordinarily be reported immediately.

2. The volume of extremely low domestic flow rates, due to leaky fixtures, faulty toilet mechanisms, etc., is surprisingly high.

Following is a table (from data used by the plaintiff's expert) that shows the result of one such study in Dallas, Texas:

Rate of flow		Water used (%)
(gal/min)	(liter/sec)	
<0.125	<0.016	9.1
0.125	0.016	4.5
0.50	0.032	1.8
1	0.063	5.0
2	0.126	11.8
4	0.232	52.4
6	0.378	14.7
10	0.631	0.7
		100.0

It was thus evident that at least 9% of the flow through the meters could be grossly underregistered.

We concluded that underregistration of the meters accounted for the overwhelming majority of the water loss and that unauthorized use and water-line leaks represented an extremely minor contribution. Therefore it appeared reasonable to predicate our estimate of leakage on actual repair records and to ignore any estimates based on unaccounted for water data.

Rainfall Infiltration

Rainfall data for the 17-yr period were derived from records of the Los Angeles County Flood Control District's rainfall gage No. 447c in nearby Carbon Canyon. See Figure 3–2 for the location of the gage in relation to Big Rock

Mesa. The Carbon Canyon gage was chosen not only because of its close geographic proximity but also because it is slightly lower in elevation than Big Rock Mesa. This fact would tend to yield conservative results, that is, slightly less rainfall than that which actually occurred on the mesa. Note that the magnitude of the isohyetal lines increases with ground elevation.

The plaintiff's expert used two other rainfall gages, which are also shown in Figure 3–2. We gave no credence to use of these gages because of their distance from the mesa and their much higher elevation.

Next it was necessary to relate the rainfall numbers to the three elements of the hydrologic cycle; namely, evapotranspiration, infiltration to the groundwater regime, and surface runoff to the ocean.

Evapotranspiration includes the process of direct evaporation to the atmosphere as well as the indirect process of transpiration, wherein trees and plants absorb water through their roots and transmit it through the plants, where it escapes through pores in the leaves. The proportion of rainfall consumed in evapotranspiration depends on a host of factors, including topography, soil porosity, type and extent of vegetation, and weather conditions such as temperature, humidity, and wind. With so many variables involved, a quantitative analysis of evapotranspiration becomes exceedingly difficult and time-consuming.

For our analysis we relied heavily on an irrigation-water-use study we had performed in the early 1980s for Pepperdine University at its Malibu campus. From the standpoint of evapotranspiration characteristics, Pepperdine is comparable to Big Rock Mesa. The campus lies about 5 mi (8 km) up the coast and just inland of the PCH, where weather, soil, topography, and vegetation (both planted and natural) are very similar to those of Big Rock Mesa. In the process of analyzing the evapotranspiration characteristics of the landscaped and natural areas we found that, on an overall average, the area would consume about 2 in. (5.1 cm) of rainfall and/or irrigation in evapotranspiration before significant infiltration and runoff occurred.

In keeping with the results of our Pepperdine study, we assumed the initial 2 in. (5.1 cm) of each storm would be consumed in evapotranspiration. A storm was arbitrarily defined as a continuous period of rainfall with no more than one intervening day of zero rainfall during the period. All storms for the 17-yr period producing in excess of 2 in. (5.1 cm) were tabulated. The excess amount represented infiltration or runoff, and the next step was to estimate the proportion of each.

The most precise approach would have been to conduct permeability tests over the entire mesa, but this would have been an exceedingly expensive undertaking that we considered not cost-effective in terms of the degree of accuracy required for the analysis. Instead we relied heavily on infiltration data compiled by the Los Angeles Flood Control District as well as our own judgment based on knowledge of the existing soil types, topography, and extent of development. Our analysis indicated an average infiltration pro-

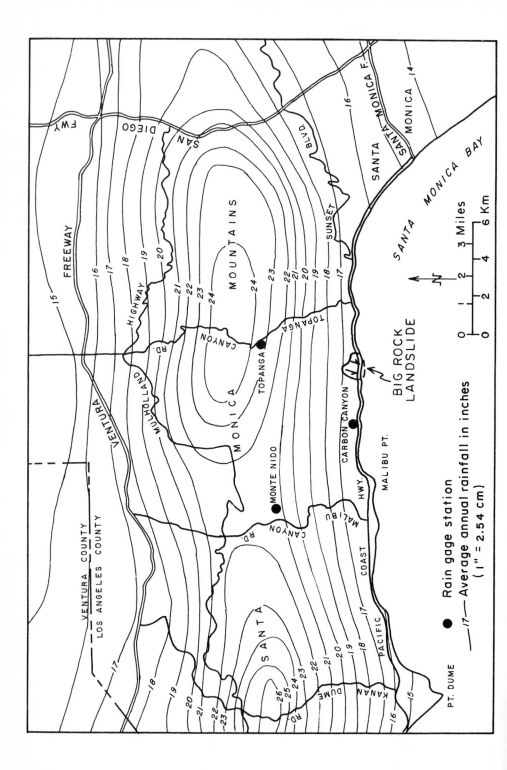

portion of 60% for all but the first 2 in. (5.1 cm) of every storm. The remaining 40% represented runoff from the unpaved portion of the mesa.

Next it was necessary to determine the surface boundaries of the landslides and to plot them on a topographic map (Fig. 3–3). Since the boundaries were then well defined by scarps, this was accomplished quite easily. The surface areas with which we were concerned for this purpose were approximately as follows:

1. Main active slide area from head scarp to top of bluff   115 acres   (47 ha)
2. Extended active slide area from headscarp to top of bluff   45 acres   (18 ha)
3. Area of tributary watershed to active slides top of bluff   26 acres   (11 ha)
         Total   187 acres   (76 ha)
4. Area of paved roads and driveways within slide and tributary water sheds draining to storm drains that removed waters from mesa   12 acre   (5 ha)
         Net area subject to infiltration   175 acre   (71 ha)

See Figure 3–4 for delineation of these data.

Of the 175 acres (71 ha) we made no distinction between natural and graded landscaped areas since the soils were generally identical. Roof areas were not deducted because roof drainage generally spilled directly onto landscaping. The tributary watershed was included because, in the absence of geologic data to suggest diversion, there was no reason to believe that underground water migration would not flow in the direction of surface contours and eventually reach the aquifer underlying the slide mass.

The total volume of infiltration was then calculated as follows:

$$V = A \times \frac{R - ET}{12} \times 0.60$$

where V = infiltration volume in acre feet
      A = 175 acres
      R = total rainfall in inches
   ET = 2 in. lost in evapotranspiration
or

$$V = A \times \frac{R - ET}{100} \times 0.60$$

FIGURE 3–2

Isohyet lines indicating average annual rainfall. Source: Los Angeles County Flood Control District.

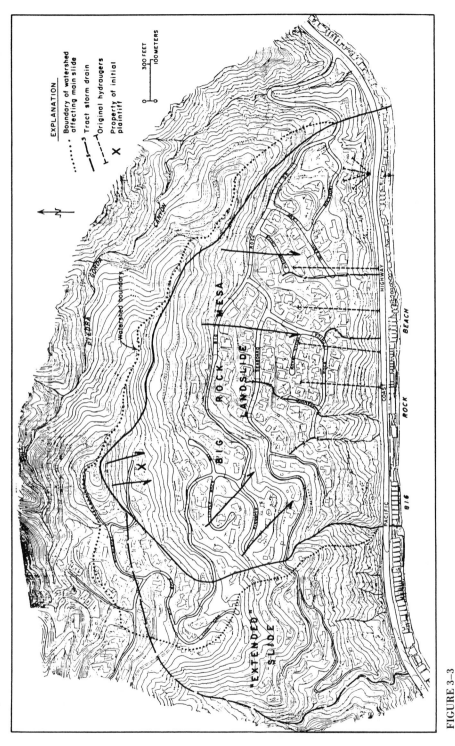

FIGURE 3–3
Site plan of the Big Rock Mesa landslide.

where V = infiltration volume in hectare meters
      A = 70 ha
      R = total rainfall in centimeters
     ET = 5.08 cm lost in evapotranspiration

Table II summarizes the infiltration computed on an annual basis.

After completing the analysis, two other particularly interesting related studies came to light. First we learned that the county had employed a consulting agronomist to prepare a detailed and comprehensive report regarding evapotranspiration on the mesa. The bottom line of this report was that 62% of all rainfall was consumed in evapotranspiration. This appeared to subtantiate solidly our 64% figure and our Pepperdine study results.

The other study was made by the plaintiff's expert and surfaced during one of his depositions. His approach to estimating infiltration was to assume that the first inch of each storm was lost in evapotranspiration, that the second inch completely infiltrated, and that all rainfall after the second inch became surface runoff.

	Percentage of total rainfall
Evapotranspiration	43
Runoff	32
Infiltration	25

One cannot attach any scientific significance to the closeness of our infiltration figures. There was, however, the practical assurance that our figure would not be attacked.

Sewage Effluent Infiltration

All residences on Big Rock Mesa possessed essentially identical private sewage disposal systems, that is, septic tanks and seepage pits. House plumbing systems discharged into the septic tanks, which separated sludge from liquid and provided crude biologic treatment. The effluent from the septic tank then flowed to a seepage pit or cesspool, from which it percolated into the ground.

We assumed that all the effluent from the system infiltrated to the groundwater regime and the ancient landslide. Reasons for this assumption were twofold: First, the seepage pits were generally 20–30 ft (6–9 m) deep, well below any tree roots; and second, construction permit applications for the septic systems necessitated percolation tests be made to assure their proper function.

Two basic methods were available to estimate the volume of effluent

TABLE II
Rainfall Infiltration Summary

Rainfall season	Total		Total for individual storms in excess of		Evapotranspiration		Runoff to ocean[a]		Infiltration to groundwater	
	(in.)	(cm)	(2 in.)	(5.1 cm)	(acre ft)	(ha m)	(acre ft)	(ha m)	(acre ft)	(ha m)
1966/67	20.75	52.7	8.28	21.03	182	22	48	6	72	9
1967/68	13.75	34.9	4.99	12.7	128	16	29	4	44	5
1968/69	23.91	60.7	12.06	30.6	173	21	70	9	106	13
1969/70	9.39	23.8	0.50	1.3	130	16	3	0	4	0
1970/71	12.00	30.5	9.60	24.4	108	13	56	7	84	10
1971/72	10.60	26.9	6.46	16.4	60	7	38	5	57	7
1972/73	18.88	48.0	3.94	10.0	218	27	23	3	34	4
1973/74	17.04	43.3	7.27	18.5	142	17	42	5	64	8
1974/75	15.95	40.5	3.65	2.3	179	22	21	3	32	4
1975/76	7.16	18.2	0.10	0.3	103	13	1	0	1	0
1976/77	13.12	33.3	2.56	6.5	154	19	15	2	22	3
1977/78	35.84	91.0	16.74	42.5	279	34	98	12	146	18
1978/79	19.21	48.8	3.76	9.6	225	28	22	3	33	4
1979/80	30.83	78.3	19.32	49.1	168	21	113	14	169	21
1980/81	10.77	27.4	1.01	2.6	142	17	6	1	9	1
1981/82	14.45	36.7	3.47	8.8	160	20	20	3	30	4
1982/83	33.38	84.8	9.44	24.0	349	43	55	7	83	10
Total	307.03	799.9			2900	357	660	81	990	122
Average annual	18.06	65.0			171	21	39	5	58	7
Percentage of total rainfall					64		14		22	

[a] From 175 acres (70 ha) only.

infiltrating to the groundwater from the septic systems. One was to calculate total water usage by means of meter records for all the homes within the slide area and contributary watershed over the 17-yr period and then to attempt to analyze the percentage of interior use versus exterior use, principally irrigation. The other method was to estimate interior usage predicated on records of studies that had been conducted in comparable communities. We chose the latter approach because assembling meter records for each of the homes for 17 yr would have been tedious and time-consuming and, in the end, we still would have had to estimate the proportion of interior use.

Interior water consumption varies with the size of dwelling, number of plumbing fixtures, number of inhabitants, whether single-family or multiple units, and last but not least the level of affluence of the household. As a rule, families in higher economic strata tend to use more water per capita. This does not imply that they are cleaner, only that they generally possess more water-consuming fixtures, waste more water, and have more nonresident water users such as maids and guests.

A study in nearby Santa Monica, published by the American Society of Civil Engineers in 1974, indicated average per capita residential interior water use of 92 gal (349 liter) per day. Santa Monica represents a wide spectrum of single-family homes, apartments, and condominiums ranging from low to high income. Compared with the predominantly high-priced single-family homes in Big Rock Mesa, we believed 92 gal (349 liter) per day was low for Big Rock. Other studies of more comparable single-family areas in Palos Verdes and Orange County revealed interior usage of over 100 gal (379 liter) per day. We were convinced that each mesa resident was using at least 100 gal (379 liter) per day.

To determine the average number of residents per dwelling we consulted census data for the time period as compiled by the Los Angeles County Planning Department. Our interpretation of the census data resulted in a figure of about 3.25 persons per dwelling. Thus the average interior household use was 325 gal (1230 liter) per day.

At this point it was interesting to speculate as to the plaintiff's attorneys' strategy relative to septic system infiltration. If they successfully downplayed its magnitude, infiltration from water-line leaks would assume relatively greater significance. On the other hand, testimony from the defendant's experts might convince the court that, without septic system infiltration, the landslide would not have occurred. Under this circumstance the plaintiff's lawyers might successfully argue that the county must bear the responsibility for allowing development without sewers. It was not long before we learned of the plaintiff's position. The plaintiff's expert testified in his deposition that he did not think the interior per capita usage was greater than 65 gal (246 liter) per day. To verify this belief he had selected 30 homes on a random basis and checked their water meter records for the two wettest months in recent history, January and February 1983. He reasoned that hardly anyone would knowingly irrigate during this period and, therefore, the metered volume would reflect almost entirely internal use. This approach made a lot of sense. His calculations resulted in an average of 156 gal (591 liter) per dwelling per day. On the basis of his census study he assumed an average of 2.4 persons per dwelling. Thus, 156/2.4 = 65 gal (246 liter) per capita per day neatly substantiated his testimony. However, under deposition, the defense attorney found that in arriving at 156 the expert had mistakenly divided the total 2-mo meter readings by 120 day instead of 60 day to arrive at average daily water consumption. Thus all his subsequent figures were 100% low. If his census analysis had been accurate, the daily per capita interior use would have been 130 gal (492 liter) per day.

After checking the meter records and his mathematics we decided to use the basic data produced by the plaintiff's expert. In effect we accepted 312 gal (1180 liter) per dwelling per day as reasonable. With our census figure of 3.25 persons per household the per capita daily use would be 96 gal (364 liter), which was very close to the figure of 100 gal (379 liter) we had decided to use.

We next researched Los Angeles County Building Department and County Assessor records to ascertain how many homes actually existed within the landslide and tributary areas for each of the years from 1967 to 1983. Table III summarizes the volume of sewage effluent in acre feet for each year. Because of the close correlation between the figures for average household interior use of the plaintiff's expert and our own, we felt the volumes to be reasonably accurate.

Infiltration Due to Irrigation

The waterworks district had calculated, by means of its individual meter records, that the average daily water consumption for the 216 dwellings on the mesa was 130,000 gal (492,700 liter). The exterior, or irrigation, usage was therefore 130,000 − 312 × 216 = 62,600 gal (237,250 liter) per day.

Average usage was thus split nearly in half between interior and exterior. However, there is a human tendency to overwater, particularly in affluent areas, probably owing to a combination of carelessness and automatic sprinklers, for which the timing and duration are not adjusted often enough. Experience with this problem on several projects, including the aforementioned Pepperdine study, has shown that the installation of sensoring devices, which automatically close sprinkler valves when maximum evapotranspiration consumption has been realized, results in a 15–20% reduction of irrigation water use.

These data formed the basis for assuming that about 10% of all irrigation water on the mesa infiltrated to the groundwater regime and that about 7% ran off into the streets and storm drains.

The following equation shows the average annual infiltration of irriga-

TABLE III

Sewage Effluent from Big Rock Mesa for the Period 1967–1983 Based on Average Use of 312 Gal per Day per Dwelling

Year	Approximate number of dwellings	Annual volume sewage effluent		Cumulative	
		(acre ft)	(ha m)	(acre ft)	(ha m)
1967	152	53.0	6.5	53.0	6.5
1968	165	57.5	7.0	110.5	13.6
1969	178	62.0	7.6	172.5	21.2
1970	182	63.4	7.8	235.9	29.0
1971	186	64.8	8.0	300.7	37.0
1972	190	66.2	8.1	366.9	45.1
1973	194	67.6	8.3	434.5	53.4
1974	198	69.0	8.5	503.5	61.9
1975	202	70.4	8.7	573.9	70.6
1976	206	71.8	8.8	645.7	79.4
1977	209	72.9	9.0	718.6	88.4
1978	212	73.9	9.1	792.5	97.5
1979	214	74.6	9.2	867.1	106.7
1980	216	75.3	9.3	942.4	116.0
1981	216	75.3	9.3	1017.7	125.3
1982	216	75.3	9.3	1093.0	134.6
1983	216	75.3	9.3	1168.3	143.9
Total		1168.3	143.9		

Average annual = 68.7 acre ft.   (8.5 ha m)

tion water on the mesa for 216 homes: $10\% \times 62{,}600$ gal $\times 365$ day $= 2{,}285{,}000$ gal or 7.0 acre ft   (0.86 ha m). For the entire 17-yr period, adjusted to reflect the actual annual number of homes, the volume of infiltration would have been 100.4 acre ft (12.4 ha m). This figure was never challenged by the plaintiff's attorneys.

Infiltration Summary

Total infiltration for the 17-yr period was summarized as follows:

	(acre ft)	(ha m)	Percentage of total
Rainfall	990.0	121.8	43.77
Sewage effluent	1168.3	143.7	51.65
Irrigation	100.4	12.3	4.44
Water-line leaks	3.1	0.4	0.14
Total	2261.8	278.2	100.00

Aside from legal concepts of "substantial contributing cause" it was apparent that, if our analysis of the various sources of infiltration was reasonably accurate, water-line leaks constituted an insignificant contribution to groundwater.

Another interesting aspect of the analysis is that total effluent infiltration from the septic system actually exceeded total rainfall infiltration. Figure 3–4, which graphically depicts annual contributions from all sources, reveals

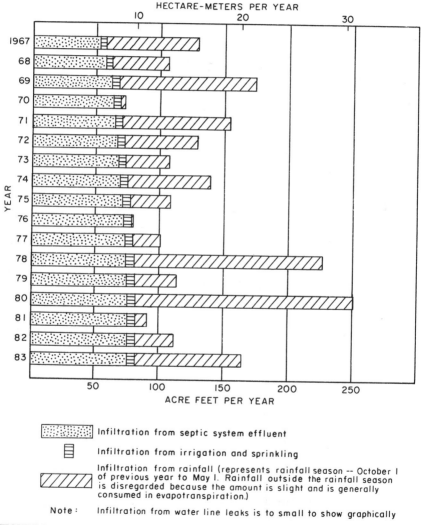

FIGURE 3–4

Infiltration summary chart.

the contrast between the relatively steady contribution from effluent and the erratic contribution from rainfall. Whereas all sources contributed to the general rise in groundwater levels, it was evident that the abnormally high rainfall that occurred in 1978, 1980, and 1983 had a triggering effect on the slide. This is not to say that the slide would have been averted had the rainfall infiltration not peaked during those years, rather merely that the timing may have been different.

Effect of Site Improvements

Another facet of our hydrologic investigation was prompted by the fact that the plaintiff's attorneys were bringing suit against the county for inverse condemnation, claiming, in effect, that improvements that were dedicated to or built by the county became a substantial cause of the landslide.

Such improvements included the streets, storm drainage facilities, and water distribution systems. The analysis of water-main leakage had been completed, and the court would have to decide whether our methods and results were reasonable and, if so, whether they constituted substantial causation. To determine the effects of streets and storm drains on infiltration we calculated what the evapotranspiration, runoff, and infiltration would have been over the same 17-yr period had there been no development at all on Big Rock Mesa.

We found that the streets and storm drains had a *net* beneficial effect as far as decreasing infiltration to the groundwater regime was concerned. Although there were some instances where development concentrated runoff, in general the streets intercepted runoff and conveyed it to storm drains, which, in turn, discharged it to the ocean.

From a legal standpoint, to prove inverse condemnation one must show that the *net* effect of public improvements was detrimental in terms of causing damage to private property. Our analysis indicated that this was certainly not true as far as the streets and storm drains were concerned.

It was interesting that, in the matter of damage caused by septic system infiltration, the plaintiff's attorneys attempted to make a case for inverse condemnation against the county waterworks district by means of this logic:

1. The waterworks district obtained most of its water supply from local wells.
2. The district then delivered the water to Big Rock Mesa residents in its distribution system.
3. The residents consumed the water and discharged it through their septic tanks into their seepage pits.
4. At least some of the effluent seepage was naturally filtered and eventually extracted from district wells.

5. The cycle was complete and the district had never really relinquished ownership of the water along its route.

Therefore, the rise of the water table attributable to effluent seepage was caused by the district's water. Under this seemingly unique and creative legal theory the homeowners were merely renting the water.

Water Budget

Water budget is a simple concept of equating groundwater inflow to groundwater outflow, with the excess of inflow over outflow resulting in a rise in the water surface. It was hoped that the groundwater hydrologist could establish a water budget to determine the fluctuations in Big Rock Mesa groundwater levels that occurred after development.

Of the two elements of inflow, infiltration and migration, infiltration had already been estimated. From inspection of the topography (there was a deep canyon uphill from the mesa) it was apparent that nearly all groundwater migration toward the mesa would be intercepted by the canyon.

However, migration *from* the mesa proved difficult to estimate on any theoretic basis, mainly because of the extreme variations in the permeability of the underlying rock strata. Borings could have been made to test permeability, but such testing did not appear to be cost-effective. Besides, it was mutually agreed upon among the consultants that there were enough water-level records from water wells to establish a fluctuation pattern.

Geologic Investigation Principles

Basically, in nearly all forensic investigations of landslides, including Big Rock Mesa, the geologists are the professionals who eventually determine the geometry and causes of the slide and thus become the pivotal expert witnesses of litigation issues. The geotechnical engineers and hydrologists provide valuable calculations and test data, but it is the geologists who must ultimately interpret these data in relation to their own findings to arrive at the most reasonable hypotheses and opinions in connection with the slide, that is, (a) why it happened at all; (b) why it happened when it did; (c) its geometric configuration; (d) the prognosis for expansion of the slide if no mitigation measures are undertaken; and (e) what mitigation techniques, if any, are most appropriate.

Essentially there are two ways the forensic geologist can direct his or her investigations. Unfortunately the prevailing methodology is for the attorney to suggest (or dictate) a conclusion or conclusions that would most closely conform to an established legal theory or to case law. The geologist then works backward to evolve a hypothesis that would best fit the desired results. To use this process is both scientifically unsound and professionally unethical. The proper scientific approach involves the use of multiple working hypotheses untainted by legal considerations. Moreover, investigating all possible hypotheses can benefit the geologist's attorney by providing potential rebuttal material for use against opposing expert witnesses.

The method of multiple working hypotheses is simply the process whereby all of the possible explanations of a given phenomenon are identified and evaluated. The valid explanation is the one that passes all critical tests.

Conducting a scientific inquiry can be inductive, deductive, or probabilistic. Inductive reasoning involves inferences from empirical data, whereas inferences in deductive reasoning are based on general principles without reference to empirical data. Probabilistic induction involves inference of the likelihood of a cause or event, with such probabilities determined from particular data. A geologic forensic investigation should involve only inductive and/or probabilistic reasoning.

The initial considerations in such an investigation are twofold: What factual information is known, and what are the current thoughts (speculations, hypotheses, theories) regarding the event? The primary empirical data must be kept separate from the interpretations of the data so that interpretations are not used as observations. Specification of the event and self-education about the existing information pertaining to the event put the investigator in a position to formulate an array of initial hypotheses.

At this stage the geologist begins to collect new data from site inspections, conduct preliminary analyses of soil and rock samples, and interview eyewitnesses to the event, etc. The initial hypotheses can then be reformulated to a point where the investigator is able to specify the work needed to be done to test all the hypotheses. This work program is then carefully coordinated with and approved by the attorney regarding content, scope, cost, and time. It must be recognized that the process of formulating the scope of work will usually require some compromise between the geologist and the attorney. Factors the attorney must consider are the dollar amount of the lawsuit, the importance of the outcome of the litigation as precedent for future related cases, the time available within the litigation schedule, and the likely quality and scope of opposing expert testimony. In particularly important cases the geologist may be given carte blanche to do everything he or she deems necessary to fully test every hypothesis. The attorney may even suggest an expansion of the work scope. In most cases, however, there are financial and time constraints that may limit the scope of the investigations. Usually such constraints can be resolved in a manner that will not destroy the essence of the multiple working hypothesis method.

Once approved by the attorney, the investigative phase might include such work as a geomorphic history of the general area of the event, research of past topographic maps and surveys, geologic mapping of the slide and surrounding area, study of previous geologic reports and aerial photographs, authorization of additional photogrammetry, and test borings and excavations. The data generated permit the application of multiple working hypotheses as follows:

1. The hypotheses that might be able to explain the observations are specified.
2. The new data are processed and organized.
3. The new data are analyzed and interpreted.
4. The hypotheses are evaluated and conditionally accepted, modified, or rejected.

A new set of hypotheses is thus generated, and the process is repeated until the number of acceptable hypotheses is at a minimum. Ideally only one hypothesis will survive multiple tests, and that hypothesis is considered the valid one through the exclusion of all other possibilities.

In practice the process may not prove to be this neat and clean. The end result of a geologic scientific inquiry may merely be to favor one hypothesis over others. Unfortunately the term "favored hypothesis" is not popular in the courtroom, where the expert witness is generally expected to render an unqualified professional opinion. At first glance the distinction seems picayune. However, an opinion based on a favored hypothesis does not rule out the possibility that it may not be the only valid hypothesis. Using such an expression on the witness stand could impart to the judge or jury the feeling that the expert is tentative and lacks confidence in his or her testimony. However, given the increasing sophistication of judges and jurors, such a declaration might actually enhance the credibility, honesty, and scientific integrity of the expert in the eyes of the court. We would hope, possibly naively so, that this would be the case. Nevertheless, we feel that the method of multiple hypotheses is the sole professional approach to scientific truth and integrity in geoscience. Failure to use it can only backfire in the courtroom and damage the professional reputation of the expert as well.

Historic Development of the Landslide

An understanding of the structural evolution of the Santa Monica Mountains is relevant in trying to understand the development of the Big Rock Mesa landslide. The highland area in the vicinity of the landslide was composed of young sedimentary rock that had been pervasively fractured and faulted before the landslide developed. Slopes composed of such bedrock are likely to be weak and susceptible to failure. It is possible that favorably oriented segments of existing fault surfaces were reused as part of the failure surfaces beneath the landslide.

The slip surface of a landslide is, by definition, a fault. In fact a landslide prism is, by definition, a micro–fault block, albeit a small one that is usually displaced on all sides from its original position. It is often difficult to distinguish between a shear zone created by a landslide from a shear zone created during other types of faulting. Unscrupulous field geologists have been known to map all shear zones as faults (even ones that may have been formed at the base of landslides) and then report that there is no evidence for recent displacements along the shear zone (even when there is no evidence against recent displacements either). Fault gouges, slickensides, slickenfibers, and all of the other attributes of fault zones can be observed within the shear zones produced by landslides.

Groundwater flow was locally impeded by the low permeability of clay-rich gouge along the faults, while elsewhere the rate of groundwater flow was enhanced by secondary or fracture permeability. Fractures that were exposed at the surface increased the efficiency with which rainwater could percolate into the subsurface. The fractured and faulted character of the bedrock inevitably resulted in an inhomogeneous, discontinuous series of confined, unconfined, and perched aquifers.

Earthquake activity and the elevation of mountain ranges along thrust faults are inextricably linked. A recent map of earthquakes that occurred in Southern California from 1980 through 1986, including microearthquakes with magnitudes too small to be felt by humans at the surface, locates hundreds of events in the vicinity of the Malibu coast fault, both onshore to the north and offshore to the south. Prehistoric earthquakes undoubtedly accompanied the folding and faulting that elevated the Santa Monica Mountains. The sudden release of seismic energy in an earthquake can cause the ground surface to lurch and be broken by extensional and shear fractures. This fracturing decreases the bulk strength of rock units, creates secondary permeability, and enhances the infiltration of surface water into the subsurface. Hillside areas are therefore somewhat more susceptible to failure after earthquake events, particularly when they occur during rainy seasons.

Mean sea level has fluctuated widely during the last 2 million years owing to advances and retreats of glacial ice. The most recent glacial maximum was approximately 18,000 yr ago, at which time sea level was perhaps 325–425 ft (100–130 m) lower than it is today. If all of the ice on Earth were to melt, mean sea level would be approximately 215 ft (65 m) higher than it is today. Hence the total variation is thought to be just under 650 ft (195 m). Sea-level variations in the past 18,000 yr, since the great continental ice sheets retreated to the polar regions, have probably been relatively small. A small glacial advance from the 1400s through the mid-1800s probably resulted in a lowering of sea level. Sea level is currently thought to be rising, a trend that will continue if predictions regarding global warming are valid.

Rising sea level can dramatically affect coastal highland areas, particularly where they are not protected within bays. Wave action is an effective and rapid agent of erosion and transportation of sediment onto the continental shelf. As sea level rises, slopes facing the surf zone are undercut and over-steepened by surficial failures to the point that they develop the characteristically steep coastal seacliffs and bluffs. These steep faces are gravitationally unstable because of their high relief, tending to fail and to expose new slope areas to erosion. Seaward of the surf zone a flat and nearly horizontal wave-cut bench is developed. The sharp corner at the extreme topographic step between the wave-cut bench and the steep seacliffs provides a stress concentrator that localizes the toe of landslides at the base of the slope.

One curious attribute of the Big Rock Mesa landslide is the apparent presence of an active slip surface about 200 ft (60 m), below the level of the PCH, which runs along the bottom of the seacliff below Big Rock Mesa. The depth of this slip surface was eventually confirmed by means of a boring at the PCH. In addition, there is evidence that this slip surface extends approximately 1000–1200 ft (300–360 m) offshore. This possibility is indicated by the results of a shallow-marine seismic survey, the reported presence of a subdued topographic ridge at that location, and the observation of plumes of muddy seawater offshore, where the landslide toe might be occasionally rising above the surrounding seafloor. A hypothesis for interpretation of the development of this configuration is shown in the accompanying figure: (*a*) Early rotational motion of the landslide, at a time of lower sea level, established a deep slip surface. (*b*) Bluff erosion concurrent with later motion of the slide mass caused the volume of the landslide prism to be decreased and part of the slip surface to be stranded. Today the active landslide is smaller than the original slide and uses both the original slip surface and the more recent slip surfaces during periods of activity.

Pleistocene marine terraces have been mapped as high as 4000 ft (1200 m)

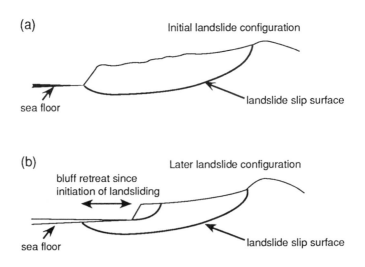

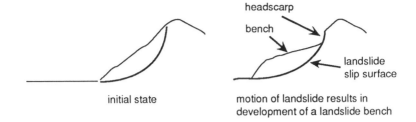

initial state

headscarp
bench
landslide
slip surface

motion of landslide results in
development of a landslide bench

above present sea level in the Malibu area. Quaternary nonmarine terraces associated with the coastal environment have been mapped at elevations of up to approximately 700 ft (210 m) in this same area. Marine terraces at high elevations in the Santa Monica Mountains are sometimes interpreted as evidence that sea level was once as high as the current elevation of the marine terrace. The interpretation is incorrect because it does not account for the continuing uplift of the range, which elevates the marine terraces relative to an observation point located on stable ground outside of the mountain range. Mountain uplift and sea-level fluctuations act concurrently to produce an array of coastal terraces at a variety of elevations in the Santa Monica Mountains.

The motion of rotational landslides can also create a terrace-like geomorphic bench below a ridge line. Often these landslide benches lie within arcuate depressions formed by the headscarp of the landslide. A wave-cut bench can also be elevated by landslide motion if it lies along the toe of a landslide. At Big Rock Mesa this mechanism for terrace uplift is well displayed by the uplift of the PCH across the toe of the landslide.

Geomorphic Investigation

This investigation was primarily accomplished by means of field reconnaissance, interpretation of aerial photographs (Fig. 3–5), and review of previous reports. The purpose of the study was to refine the latest (1980) USGS mapping of the area surrounding the landslide to better understand the following:

1. The extent of the active landslide on Big Rock Mesa
2. The extent of the postulated ancient landslide encompassing the currently active Big Rock Mesa landslides
3. The location of active yet unmapped landslides in the vicinity of Big Rock Mesa
4. The location of slope areas that are geomorphically indicative of prior slope failure

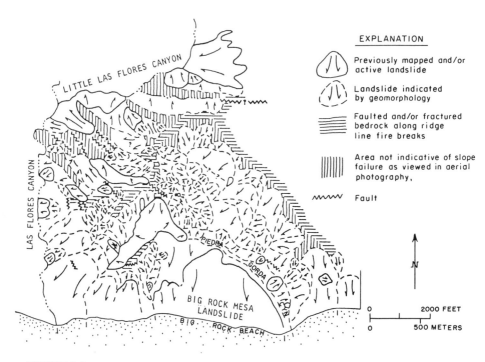

FIGURE 3–5
Photo-interpretation map of landslides in vicinity of Big Rock Mesa.

5. The character and distribution of local faults
6. The structural and lithologic character of local rock units

Field investigations were conducted in the area bounded by Little Las Flores Canyon on the north, Big Rock Beach on the south, Las Flores Canyon on the west, and the ridge between Piedra Gorda and Pena Canyons on the east. This is an area of roughly 2.25 mi^2 (5.85 km^2), with a maximum relief of approximately 1600 ft (480 m) (see Fig. 3–3).

The coastal mountains in the Malibu area have been known to be highly susceptible to slope failure for many years as documented in USGS studies. Mapping by the USGS has been conducted in the area on an irregular basis since 1960, culminating in the publication "Geologic Map of East-Central Santa Monica Mountains, Los Angeles County, California." The USGS is characteristically very conservative in mapping landslides; an area that may be very indicative of landsliding on the basis of geomorphology may not be mapped as a landslide without further physical evidence of slope failure. The unequivocal establishment of a slope feature as a landslide is often too costly an undertaking to be reasonable in general geologic reconnaissance mapping, although it is critical to the type of site- or tract-specific geologic mapping that precedes the development of an area for residential or industrial use. It is therefore significant that as much as 20% of the Big Rock Mesa area was identifiable as being involved in landsliding by the conservative mapping policies of the USGS.

Additional recently active landslides have been mapped by a number of geologists, including private consultants and geologists employed by Los Angeles County. A very conservative estimate is that 25% of Malibu in the Big Rock Mesa area is known to be actively involved in landsliding. The evidence for active landsliding included the following: damage to man-made structures and roadways; development of surface cracks; measured rotations in tilt meters; elevation changes; development of scarps; generation of mud flows, debris flows, rockfall hazards, etc.; anomalous groundwater behavior; and disruption of vegetation. The area can certainly be described as one in which a relatively high percentage of the surface is covered in landslides and in which there is a very high susceptibility to renewed landslide activity.

Three broad classes of anomalous features were relied on to discern probable landslides visible on the several sets of aerial photographs that were examined. The features thought to be indicative of landsliding included anomalous topography, anomalous drainage, and anomalous vegetation. *Anomalous topography* is evinced by the following: interrupted lineaments; "stairstep" ridge lines; lobate topographic elements on slopes; hummocky topography; arcuate, amphitheater-like topography at or below a ridge line— indicative of a headscarp; benches below ridge lines—broad, subdued topography on a slope where terrain is otherwise moderate to steep; pressure ridges; and scarps.

Indications of *anomalous drainage* include displaced drainage course around the toe of a lobate slope; linked drainages on either side of a secondary (spur) ridge or near the nose of a ridge, particularly where the intersection of the drainages with the ridge line is coincident with a topographic break in the ridge line; drainages developed in incompetent materials, as in a fractured landslide mass; and curvilinear drainage courses in an area typified by a structurally controlled trellis drainage pattern.

Patchy vegetation and unusual grassy areas are *anomalous vegetation* conditions that are visible in aerial photography.

In addition to those indicators used in the analysis of aerial photographs, several others were used in the course of the field work to identify probable landslides in the vicinity of Big Rock Mesa. These included pervasively fractured rock units in outcrops; surface cracks (e.g., transverse, radial, or "pull-apart" cracks); slickensides, gouge, and other indicators of mass movement that are not confined to a discrete fault zone; springs or seeps that are seemingly unrelated to faulting; rotated or bent vegetation; rotated blocks of rock or unconsolidated material; vegetation indicative of anomalously high water content (phreatophytes); and damage to man-made structures—tilting of fences, posts, etc.

Variations in rock type did not appear to be a controlling factor relative to the location of landslides in the Big Rock Mesa area, although lithology has been cited as a significant controlling factor in landsliding elsewhere in coastal California. The Big Rock Mesa area contains rock ascribed to the formations shown in Figure 3–6.

Nearly all of the rock encountered in the Big Rock Mesa area was pervasively fractured, if not pulverized, into fault breccia or gouge. Along the PCH on the steep slope below Big Rock Mesa, large blocks of relatively unfractured Sespe Formation sandstone and conglomerate could be seen immediately above slide surfaces, signifying that they were being carried along within the landslide mass. Relatively unfractured Sespe Formation sandstone could also be seen as triangular prisms between strands of the Las Flores thrust fault in an old cut face immediately below the ridge line that lies between Las Flores Mesa and Big Rock Mesa (Malibu Beach 7.5-minute quadrangle). The Topanga Formation sandstone, which crops out at the top of Big Rock Mesa along the active headscarp, appeared to be relatively unfractured at first glance but on further inspection was seen to be gridded with fractures at a frequency of several fractures per foot.

The fire road and adjacent fire break that runs from the water towers north of Big Rock Mesa (Topanga 7.5-minute quadrangle) west to the level ridge terminus 1400 ft (420 m) northeast of the Las Flores fire station (Malibu Beach 7.5-minute quadrangle) exposed a series of extensively fractured strata, frequently scored by diffuse zones of varicolored pulverized rock. The most common colorations in these (presumed) fault zones were green, maroon, white, and black. The diffuse character of these zones made it impossible to define a unique fault attitude at the outcrop. Some relatively unfractured

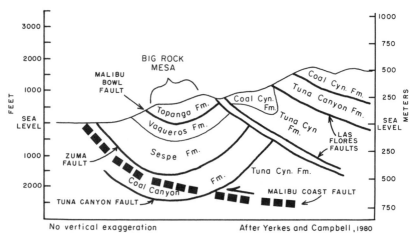

FIGURE 3–6
Structural cross section through Big Rock Mesa and vicinity.

fossiliferous sandstone bodies were encountered in places along the ridge; however, even these resistant beds were jointed into boulder-size blocks. The bedrock along the entire ridge is pervasively faulted and/or fractured. Similar relationships were noted along all of the ridges that were traversed.

The surfaces on either side of most of the fractures that were examined contained slickensides and/or a weathering rind. Some of the fractures were filled with calcite, gypsum, or a loose, dry powder presumably derived from pulverized rock. Near the surface, plant roots were pervasive within the fracture. There has probably been some development of clay minerals within the fracture systems as a result of both weathering and crushing during movement.

The joints, fractures, slickensides, and pulverized rock materials may be the result of either tectonic dislocation/deformation or landsliding, or both. Whatever the cause may be in a given location, the physical result is the same: the rock units have been structurally weakened and are highly susceptible to failure as a result of either a seismic event or landsliding, or both. To state it another way, the fractured nature of the rock units is the critical fact. Whether the pervasive fracturing was originally the result of ancient landsliding or tectonic deformation is relatively less significant with respect to present and future landslide susceptibility.

To summarize our study, Big Rock Mesa and vicinity is extensively involved in landsliding, with at least 89% of the surface area covered by known or suspected landslides. The remainder of the area is pervasively fractured, rendering it highly susceptible to failure.

Landslides in the Big Rock Mesa area are characteristically compound in nature: within the large host mass several smaller landslides exist whose slide planes are wholly enclosed within the larger slide mass. The smaller, subsidiary or parasite landslides probably move more frequently, more rapidly, and with greater surficial disruption than does the larger slide mass. Multiple slide planes within the large landslide mass resulted in the entrapment of groundwater within prisms of landslide material bounded by the many slide planes because of the formation of impermeable clay gouge along failure surfaces. As a result, the smaller, more surficial landslides within a compound landslide mass became saturated much more rapidly owing to natural and artificial input of water, which in turn resulted in the renewed downslope movement of the slide mass. Progressive saturation, if allowed, would lead to movement along progressively deeper slide planes and/or slide planes that would progress upslope.

Application of Multiple Hypotheses Method

Two sets of multiple hypotheses were generated in connection with the Big Rock Mesa landslide. First, when development was being planned in the early 1960s, two hypotheses emerged from studies of the origins of the

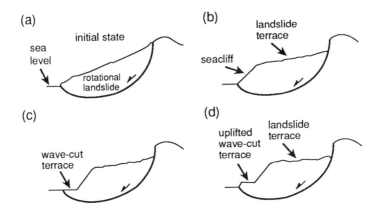

amphitheater-shaped geologic feature of the mesa. Second, when precursor signs of the slide appeared in the mid- to late 1970s, various hypotheses developed about its size and shape, whether it was single or compound, and how it started and progressed.

The early 1960s interpretations of the geologic features were of largely academic interest to us since, by the time we were employed, everyone was in agreement that the feature was indicative of past landslide activity and not merely a marine terrace. In 1962 the marine terrace hypothesis was accepted by everyone except the author, who as discussed previously was discharged as a consultant to the developer. Still, it is instructive to pursue the reasons for the failure of the geologists, both in the private and the public sectors, to recognize the landslide.

How can a geoscientist tell the difference between a wave-cut marine terrace or a stream-cut nonmarine terrace and a landslide terrace? The presence of a thin veneer of marine or river deposits on top of the bench is evidence that it was, at some prior time, at the bottom of a stream or below sea level; however, this is not necessarily evidence that the bench is not also atop a landslide. Porous and permeable terrace deposits, whether marine or nonmarine, facilitate infiltration of rainwater and surface runoff from higher slope areas. The more gentle slope angle of the terrace tends to slow the flow of runoff water, which also facilitates infiltration. Hence terraces along hillsides tend to act like drains for rainfall and surface runoff.

The typically high level of the water table in the area of a drainage terrace maximizes the weight of the slope, thereby maximizing the driving force responsible for slope failure. High pore fluid pressure along existing or potential failure surfaces below the terrace acts to reduce the resisting forces that tend to keep the slope materials in place. In short, it may not be particularly important to differentiate between landslide terraces cut by stream or marine wave action. The existence of a terrace within a hill slope, regardless of the origin of the terrace, should be considered an indication of the high potential

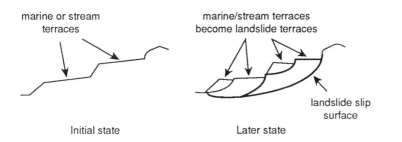

marine or stream terraces

Initial state

marine/stream terraces become landslide terraces

landslide slip surface

Later state

for the development of reactivation of landslides along that slope. Marine and stream terraces cannot be considered intrinsically safe to develop.

In the case of Big Rock Mesa it seems that great significance was ascribed to the relatively intact condition of the sedimentary rock along the seacliff below the mesa. Where sheared rock was observed, the shearing was attributed to faulting related to the uplift of the Santa Monica Mountains and not to faulting (normal faulting and landsliding) related to the erosion and gravitational collapse of the range. The condition of the rock in the interior of a large landslide depends on many factors, including the rate at which the landslide has moved (strain rate), the distance from the rock under observation and the slip surfaces, the strength and material properties (rheology) of the material between the observed rock and the slip surfaces, the strength of the observed rock, the pore fluid content (percent saturation) and chemistry, the effects of chemical weathering, and the presence and geometry of preexisting faults and fractures. Although landslide prisms tend to deform during landslide motion, volumes of rock within the landslide prism may move as a relatively undeformed plug surrounded by more extensively sheared material. It is not valid to interpret the presence of local, relatively intact rock as conclusive evidence that the rock has not been involved in landsliding. It is often only an undeformed block riding within the landslide.

When the author observed the sheared rock in the seacliff in 1962 he could not be certain the deformity was due to uplift faulting or landslide activity. However, the fact that the adjacent seacliff rock formations, both to the east and to the west, were relatively intact gave credence to the landslide hypothesis. A deep test boring at that time could probably have proved the hypothesis. However, the county wanted no reference to a questionable landslide in the author's report unless he could provide proof of its existence. And the developer, already enraged that the subject was even discussed, was certainly not about to pay for a boring.

By the mid-1970s a sufficient pattern of cracks had appeared on the mesa to present ample evidence of landslide activity. At that time multiple hypotheses began to develop about the origin, nature, and characteristics of the slide or slides. Some of these hypotheses were unbiased and based on scientific principle. Others were primarily self-serving attempts to gain legal leverage or support the position of a client.

The hypothesis that no landslide existed prevailed throughout the 1970s and up to the time the first home was destroyed in 1983. Poor-quality fill was being blamed as the culprit causing the cracks and other signs of distress that were appearing in the ground surface and in structures at the northerly end of the mesa. As mentioned earlier the first house destroyed was the last constructed; it was completed in 1980 under a building permit issued in 1978. The geology report prepared by the owner's geologist and approved by the county indicated that there was no landslide and that the site was stable and safe for the intended construction.

Once the first house was torn apart and a scarp 6 ft (1.8 m) high developed north of the house, all consultants and county personnel belatedly conceded that there was, indeed, landslide motion occurring. At this point four primary hypotheses evolved relative to the nature of the slide:

1. It was relatively small and localized at the northern end of the mesa. This hypothesis was favored by the consulting experts for the defendant, CalTrans. If a convincing argument could be made that the slide was far enough removed from the PCH to be affected by grading of the bluff, the state would probably be exonerated, at least as far as the first plaintiff was concerned. This hypothesis was predicated on the reasoning that, at that time, there were no major signs of distress in the areas adjacent to the bluff. As time passed the bluff area became impacted and this hypothesis quickly vanished.

2. It was a single slide that started at the bluff and progressed to the north. This hypothesis was advanced by the plaintiff's experts and neatly conformed to the plaintiff's allegations against the state, to wit: that the state graded the PCH and disturbed the bluffs in the late 1920s when it assumed title to the road as a state highway. Such grading allegedly occurred in the process of straightening its alignment. This hypothesis also assumed that the state, through the years, had removed slide debris from the roadway many times as a result of bluff sloughing, thereby removing lateral support and making the bluff more susceptible to future failure. The hypothesis of a single slide starting at the bluff was based on the theory that movement of the bluff redistributed stresses within the bedrock in a progressive fashion until the entire mesa had been affected to some degree. The progressive stress redistribution coupled with the rising groundwater created a situation in which the delicate stability balance was disrupted and the slide put into motion.

3. Instead of a single slide, the geomorphology indicated the existence of three major slide blocks. Specific indications included topographic expression of three separate mesa-like features, three surface drainage patterns, three separate groundwater regimes, and different kinematic (direction of motion) patterns for each of the three slide blocks.

4. The parent slide was 0.50–0.75 mi (0.8–1.2 km) in length, at least 200 ft

Shallow landslide typical of the bluff portion of the massive Big Rock Mesa landslide complex.

(60 m) deep at the PCH, and toed out in the ocean. Further, there were three parasite slide blocks moving atop the parent slide in different directions. This hypothesis was predicated on kinematic studies based on strategically placed control points that were surveyed repeatedly to determine incremental direction and magnitude of motion. The depth of the parent slide at 230 ft (69 m) was confirmed when a deep boring was made by the state at the PCH. The boring also revealed the shallower slide planes as shown in Figure 3–7.

Movement at the deepest slide plane was determined from slope indicators installed in the boring. It appeared that all previous borings were simply of insufficient depth to reach the clay gouge zone of the parent slide. Many months were lost in the forensic process because this boring was delayed so long. As for the three separate shallower slides, they were confirmed by geomorphic evidence, kinematic data based on survey control points, and recent subsurface exploration conducted by others. It was thus possible for us to establish that the deep parent slide with three slides riding on top was the most valid hypothesis. However, there continued to be divergent opinion among the other geologic and geotechnical engineering consultants regarding the nature of the landslide movement. Each appeared to take a position that would be least harmful to his client's interests.

Investigation Summary

The investigation phase of the assignment consumed nearly a year. One member of our team of consultants had been designated "captain," and he coordinated our meetings with the attorneys. Several of the early conferences included all the defense attorneys and all the defense witnesses. However, as time went on there was little exchange between consultants of the defense teams, as each defendant plotted its own legal strategies. In general, most of the contacts within the waterworks district consulting group were informal and spontaneous. We had numerous chalkboard sessions where we discussed and compared our findings and theories and then proceeded to erase the board. It is generally not wise to exchange preliminary written notations, calculations, analyses, etc., because these documents have a way of being filed and ultimately showing up in someone else's deposition. This can create an unnecessary problem for the expert in having to explain why a preliminary document has been revised or discarded and is not now valid.

In the pursuit of scientific objectivity there need not be unanimity of opinion and conclusions among the various team members. There seldom is. But there should be a closely coordinated verification of factual matter and data. For instance, team members should agree on the source of rainfall data to be used after examining all sources to determine the most reliable and relevant.

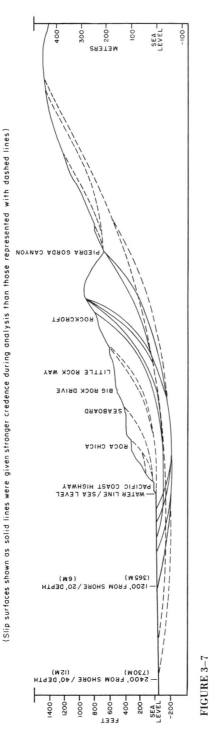

(Slip surfaces shown as solid lines were given stronger credence during analysis than those represented with dashed lines)

FIGURE 3–7

Cross section showing speculative slip surfaces of the Big Rock Mesa landslide.

In the final analysis there were several minor differences among the team members. They were, for the most part, related to how the different disciplines analyze a geologic-type failure, what data were used, and the specific method of analysis employed. Minor variables are an indicator that each discipline has followed good independent practice. There is no question in our minds that an objective team approach produces the most accurate and complete interpretations and conclusions.

The consensus among the group could be summarized as follows:

1. The landslide represented the reactivation of an ancient slide.

2. The present landslide was deep seated (about 230 ft [70 m] at the PCH), toed out in the ocean approximately 1200 ft (360 m) from the shoreline, and encompassed a total area of approximately 300 acre (120 ha).

3. At least three separate shallower parasite landslides moved in different directions atop the parent slide, each causing damage to structures above and beyond what would have occurred with the host slide alone.

4. The basic or proximate cause of the slides was an extremely high water table that began to rise after initial development in the 1960s.

5. The overwhelmingly significant sources of water contributing to the rise in groundwater levels were rainfall and domestic water from septic systems.

6. Development of the mesa had the following effects:
    (a) Relative to rainfall infiltration the *net* effect of development was to decrease slightly infiltration to groundwater. This was due to the beneficial role of the paved roads and storm drainage systems in conveying much of the runoff in impervious structures to the ocean.
    (b) The introduction of septic systems was a major cause of the rise in the water table.
    (c) Irrigation and sprinkler systems constituted a minor causation.

7. Water-line leaks from both public and private lines represented an insignificant contribution.

The net effect of all the development factors was to greatly increase water infiltration. If sewers had been installed in lieu of individual septic systems, the developed mesa would not have contributed more water to the underground regime than would an undeveloped mesa without water wells. An indirect (and arguable) result of development is that most of the water wells were abandoned when development began.

8. The system of hydraugers, which were a prerequisite of development approval, would not have been sufficiently effective in eliminating groundwater rise even if they had been properly operated and maintained. To have averted the landslides, a closely monitored water extraction program encompassing both wells and hydraugers would have been necessary from the outset of development.

9. The abnormally heavy rainfall seasons of 1977–1978, 1979–1980, and 1982–1983 played a major role in triggering the landslides and helped explain

their timing. However, even if those seasons had been normal, the slides would merely have been delayed a few years. The rise in the groundwater level was inescapable without extraction facilities.

10. Stability calculations of the parent host block, based on various groundwater-level assumptions, indicated unacceptably low or nonexistent safety factors.

Mitigation

Although studies and recommendations for stabilizing the landslide were not of direct concern in the causal-oriented litigation, they were extremely important in establishing or refuting claims for future mitigation expenses as well as evaluating the potential for further structural damage.

In general there are only a limited number of ways to stabilize a landslide. The driving force that induces landslide motion can be reduced by removing mass from the landslide. This can be done by grading the landslide mass to remove rock material or by dewatering the landslide to remove pore fluid. The net force resisting the landslide motion can be increased by buttressing the toe of the landslide or, in theory, by fusing the landslide material together and to the hillside. Fusing or cementation of a landslide is intended to increase the shear strength of the material; however, the heterogeneity of most large landslide masses makes this approach unreliable. Finally, the landslide material can be entirely excavated and the slope rebuilt in a more stable configuration.

At Big Rock Mesa several of these potential cures were immediately eliminated. The Big Rock Mesa landslide probably cannot be buttressed because of its mass, its probable extension offshore, the presence of a major highway along the toe of the landslide, and the probable eventual erosion of any toe buttress established in the nearshore environment. The landslide cannot be fused because of its great volume and its heterogeneity. The landslide cannot be excavated without destroying the properties that a mitigation effort is intended to save or preserve. Further, there would be no place to stockpile earth. Finally, even if technically feasible, the cost would exceed the value of the reconstituted real estate.

The only remaining option for stabilizing (or partly stabilizing) the Big Rock Mesa landslide is dewatering, which involves not only drainage or extraction of subsurface water but also drastic reduction of sources of groundwater: septic system discharge, rainfall infiltration, excess irrigation, and water-line leakage. Since the landslide is a compound in which several small landslide masses are clustered within the host landslide, gouge along the slip surface of secondary landslides may constitute an aquiclude, leading to multiple perched water tables. Near-surface secondary landslide prisms are therefore more likely to become saturated by wastewater and surficial infiltration, particularly during rainy seasons. Hence dewatering plans must include all landslide prisms that may be hydrologically separate within the main landslide mass. Dewatering alone probably cannot stabilize the high steep bluffs along the PCH, so it might be necessary to grade the bluff face at the expense of the first row or two of properties along the top of the shelf. This is a costly, litigation-inducing, last-resort solution.

Mitigation efforts that have been employed at Big Rock Mesa have been limited, to a great extent, to trying to control the groundwater status of the area. Attempts have been made to control surface drainage and runoff even though the surface of the landslide is locally fractured, particularly around the margins of the landslide. Water supply pipes have been laid above ground surface with expandable U-shaped sections that can extend during ground motion. Attempts have been made to establish hydraugers along the toe of the

Initial state: Unsaturated block at rest on an inclined frictional surface

Later state: Saturated block sliding on an inclined frictional surface

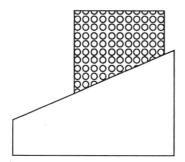

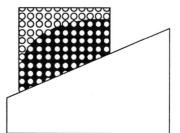

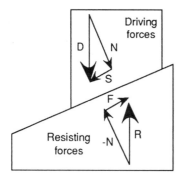

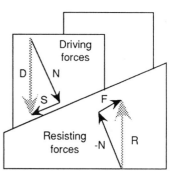

Driving forces (D) ≤ Resisting forces (R)
Shear stress at base of block (S) ≤ Friction (F)

Driving forces (D) > Resisting forces (R)
Shear stress at base of block (S) > Friction (F)

Effective normal stress at base of block (N) is reduced by pore pressure

landslide at the base of the seacliff; however, these gravity drains tend to be sheared by subsequent landslide motion and require vigilant maintenance. A system of vertical pumping wells has been established atop Big Rock Mesa, which extracts groundwater within the landslide mass and discharges it onto the roadways, into the storm drains on the mesa, and out to the ocean. Continued motion of the landslide has the potential for cracking roadways and elements of the storm drainage system.

Mitigation measures are still being conceived and implemented by the county, and an assessment district has been established for Big Rock Mesa to pay for these measures. Many residents of the district are balking at the high costs involved, particularly since a vigorous and vigilant dewatering program will have to be pursued until such time as a sewer system is functioning (if ever). Even after that it may be necessary to operate some of the dewatering wells and to monitor groundwater levels closely.

THE LITIGATION

Because of the huge number of litigants and the fact there were three major government defendants, the legal process was slow, cumbersome, and excruciatingly complicated. Fortunately nearly all of the plaintiffs were represented by only two law firms. However, there were numerous homeowners' insurance companies involved, each represented by its own law firm. If the principal lawsuit, that involving the greatest number of litigants, had ever gone to trial, it would have required a courtroom large enough to accommodate a minimum of 300 attorneys. To this end serious consideration was given to renting and refurbishing the Hollywood Palladium, a world-famous dance pavillion that once hosted the big bands of the swing era. Cost to the taxpayers for renting and remodeling this temporary courtroom was estimated to exceed $2 million. It would have been the largest courtroom in the history of jurisprudence. However, prior to signing the lease an out-of-court settlement was reached.

The strategy of the plaintiffs' attorneys was to bring a lawsuit, involving one plaintiff, to trial as quickly as possible. If they were successful in this suit they were confident the precedents established would either bode well for them in the other trials or pressure the defendants to offer a favorable settlement. There was much at stake since the aggregate of the lawsuits, including damages and emotional pain and suffering, was about $300 million.

The plaintiffs' attorneys obviously chose the initial test case property very carefully. The following factors were undoubtedly considered:

1. The house and improvements had been completely destroyed. Salvage value was minimal.
2. The lot had been rendered unbuildable except at an indeterminate cost of mitigation.

Tennis court of initial plaintiff. Within weeks the house was destroyed as well. Construction had been completed only 3 yr earlier. See Figure 3–3 for location of property of initial plaintiff.

3. The house was one of the most expensive on the mesa, about $2 million.
4. It was one of the last homes built on the mesa (completed in 1980). Thus the grading and construction complied (or at least should have complied) with the most recent stringent codes and practices.
5. The damage was obviously caused by landslide activity, not by slope failures or inadequate fill compaction.

The plaintiff's attorneys were experienced and skilled in the use of inverse condemnation as a legal concept. Under this approach they had to prove that the landslide activity that caused the damage to the plaintiff's property was the result of a government improvement, project, or policy. Negligence need not be proved in inverse condemnation, only that the project in some way damaged the private property. In the case of the state (CalTrans), the plaintiff's attorneys would argue that grading work performed along the PCH caused instability of the bluffs that, in turn, caused progressive stress redistribution, ultimately resulting in the landslide. As for the waterworks district, it would be argued that it installed, owned, and maintained the water distribution system and that water leaking from this system into the subsoil (whether within reasonable nonnegligent limits or not) constituted a claim for damages under inverse condemnation. In addition, there was the previously mentioned inverse theory wherein the district never really relinquished its responsibility through the entire use and extraction cycle.

The claims for inverse damages against the county were more subtle and would test the legal parameters of inverse claims. First, it would be argued that the county could not escape its responsibility for the hydrauger dewatering system even though it never owned or maintained the system. The reasoning would be that the county should have taken legal action against the homeowners to force continued operation or, as an alternative, taken over operation of the system and charged the homeowners for the service. Another basis for the inverse condemnation claim against the county would be that the storm drainage system, which was dedicated to the county by the developer, was inadequate to handle all the runoff from Big Rock Mesa. The key legal question in this regard seemed to be whether more water infiltrated the mesa before or after development. If the county could prove it had not worsened the situation, precedent law would not appear to penalize inadequacy.

Although not really an inverse condemnation question, the plaintiff's lawyers expounded that the development should not have been approved by the county in the absence of a sewer system. This question became a key legal issue both in the trial and in the appeal that followed. The issue went to the very heart of the controversy over whether government should be the insurer of last resort.

The Initial Trial

Our team of consultants was deposed at length by the plaintiff's attorneys, and vice versa. The civil engineering/hydrologist expert for the plaintiff was actually deposed eight times by the attorney for the waterworks district. In bulk comparison, eight volumes of depositions equals about four big-city telephone directories. At the time it seemed like a case of overkill, but, as events unfolded, it proved to be time wisely spent. We not only read those depositions, as well as all the others, but sat in on many as well. This is generally permissible if both sides agree.

The deposition is simply a questioning procedure, under oath, in which the opposition attorney asks questions of the expert concerning qualifications, results of investigative work to date, and tentative conclusions drawn. Its purpose is to help the opposing attorneys prepare their case and, at least in theory, thereby reduce trial time. Depositions were instrumental in the outcome of this case. Often a lawsuit is settled prior to trial as the result of strong testimony or evidence presented at a deposition.

The attorney on the expert's side is always present at the deposition to make certain the questions are relevant, proper, and legal. Sometimes, when objections are made, the opposing attorney will persist in trying to extract a reply, causing the expert's attorney to instruct the witness not to answer. Depositions can become as volatile and emotionally charged as a trial. In this particular case the opposing attorneys were so combative it was necessary to employ a retired judge (referred to as a "rent-a-judge") to preside over the depositions and prevent the attorneys from committing verbal if not physical abuse on one another.

Most depositions are calm and reasonable, but the expert should be prepared for the worst. The following are some hints for the expert:

- Arrangements for the deposition are generally made through the expert witness's attorney. At that time it is well to reaffirm the hourly fee rate and to find out which side is going to pay. Deposition fees are usually paid by the opposition, but not always. If the invitation to deposition is by subpoena, there should be a clear understanding that the witness is being paid as an expert by the opposing attorney. Fees should include travel time as well as living expenses if the deposition is out of town. Also, the time spent reviewing the deposition transcript should be included.
- The expert is always requested to bring to the deposition all material, including reports, notes, calculations, maps, correspondence, and billings. Prior to being deposed the expert should review the material carefully to be certain there is no client–attorney privileged information or documents included.
- At the deposition it is important to be certain each question is fully understood before answering. If a question is not understood the witness should ask for the court reporter to read it back. If it is still unclear, the deposer should be asked to rephrase the question until there is no doubt as to its meaning.
- The witness should never speculate nor guess regarding a question of fact. The most underused phrases in a deposition are, "I don't know" and "I don't remember." These replies may be embarrassing for a witness who feels that the answer ought to be known, but honest embarrassment is far preferable to an answer based on speculation that could backfire.
- The witness should pause slightly before answering complex, substantive, or exceedingly broad questions. This gives the witness's attorney an opportunity to react or object before the witness has blurted out an answer to an improper question.
- It is essential that the witness answer each question as honestly, accurately, directly, and succinctly as possible. An answer should not be expansive unless an explanation or elaboration is the only way to keep a short answer from being misleading. It is a common trait among people of science and engineering (including the authors) to want everyone to understand the complete scientific picture or theory. The deposition is not the forum for such educational activity. It should be saved for trial.

- If a "yes or no" type question cannot be answered that simply, the witness should insist on qualifying the answer. The expert is generally given a great deal more leeway to do so during deposition than at trial.
- When the expert's attorney says, "I instruct you not to answer that question," the expert should take heed. The deposer will often rant, rave, and threaten dire consequences, but the witness should hold fast. At times like this the witness wishes there were a "rent-a-judge" present.
- The expert can expect to be asked whether he or she has formulated any opinions yet as to causation. This is, of course, the key question. Answers to it should be as well thought out, definitive, and precise as possible, recognizing that the expert may face a tremendous credibility challenge if the answers change (albeit honestly so) between deposition and trial. If the expert has not yet completed the investigation and, therefore, certain answers to the questions are tentative, the expert can expect to be deposed again and again until definitive responses are forthcoming.
- Other questions always asked toward the end of the deposition are, "Have you finished your investigation?" and, if not, "What further work do you plan to do and when do you estimate your work will be complete?" The witness should be well prepared to answer these questions. Neglecting to mention a further area of investigation may cause difficulty in getting that subject admitted in court testimony. On the other hand, if there are vital investigative avenues not yet explored, the witness will probably be deposed again.
- Most important, the answers to the questions can be used against the witness at trial. Attorneys are trained to do everything within lawful means to win their case. It is not unusual for an expert witness's testimony to be impeached for errors, inconsistencies, and contradictions, even if honestly made.

There is usually ample time to painstakingly review the deposition transcript before signing. The expert should look not only for obvious court reporter or transcription errors, but at the substance of the questions and answers. There should be no hesitation in revising or expanding the answers whether because of errors, misunderstandings, or even second thoughts. (Better revising now than trying to explain at trial.) Revised answers should be reviewed with the expert's attorney so that they do not come as a surprise. Then a copy of the original corrected transcript should be obtained.

The time span between deposition and trial varies considerably. Beware when it is exceedingly short. One of the authors learned a valuable lesson the hard way in a recent case. The author had been deposed early, before completing his investigation. Then he was deposed again for two consecutive days, just two days prior to his trial testimony. His first mistake occurred at a late hour in the second day of deposition when he answered "yes" to a loaded question that should have been expanded upon. His second and more profound mistake was hurrying his review of the hastily typed manuscript and missing the first gaffe. As a result the opposing attorneys were able to successfully convince the judge that certain portions of the planned testimony be excluded. An in-chambers appeal to the judge based on the limited available time to review the deposition was to no avail. The moral is, if deposed just before trial, do not rush the transcript review.

The importance of being fully prepared for depositions cannot be overemphasized. In the Big Rock Mesa case the plaintiffs' expert made a simple but extremely pivotal arithmetic error in what was otherwise a very valid wateruse study. Through the whole series of depositions, this error, as well as several others, was never corrected. During the trial the expert, when questioned about these errors, was still confused. As a consequence he lost a great deal of credibility, even in areas of his testimony that did not contain such errors.

For the Big Rock Mesa trial we prepared many exhibits in the form of maps, cross sections, charts, aerial photographs, etc. No information was contained in these exhibits that had not been reviewed or supplied to the opposition in our deposition testimony. To do so would have risked inadmissibility at trial. At this juncture we knew that the trial would be by judge, not jury. However, this fact did not in any way affect the manner in which we prepared the exhibits or planned in general for trial. If there is any distinction between presenting testimony to a judge or to a jury, we leave such nuances to the attorneys. We have found that most juries show a great deal of sophistication, in addition to common sense, and are able to absorb complex technical testimony if presented in a suitable manner.

In preparing exhibits, consideration was given to the most effective method of presentation, such as:

- Chart and map displays, including their size, lettering, and color coding, for best visual effect from where the judge would be sitting
- Enlarged photographs (both ground and aerial) with overlays for feature identification
- Videotape accompanied by an interesting audio or in-court narrative
- Three-dimensional topographic model (which can be particularly effective, but which we did not use in this case)
- Computer-enhanced models (not used in this case)
- Last but not least the old-fashioned blackboard or easel

Another important aspect of our trial preparation was to meet at length with our client's attorneys to review the final results of our investigations and to discuss our testimony. Once the trial begins the attorneys become extremely busy, so it is wise to plan these meetings well in advance of trial.

Very few trial attorneys are schooled in the fundamentals of civil engineering or geology, but most learn quickly. We, and some of the attorneys with whom we have worked, have found it tremendously helpful to give them a crash course in those rudiments of our profession that may apply to the subject lawsuit. The process may require several sessions but, in our experience, has always been quite valuable in terms of better understanding and more enlightened communication. Our client's attorneys in this lawsuit became extremely well versed in a very short time.

It is crucial in these pretrial meetings to discuss the questions thoroughly that will be asked in direct examination, their sequence, and when and how the exhibits will be introduced. Our feeling is that completely scripted and rehearsed testimony is expensive, unprofessional, and generally ineffective. Most judges and jury members are suspicious of testimony that appears too smooth and of questions and answers that seem too pat. On the other hand, not discussing the questioning in advance can prove disastrous. One of the authors, in another trial, was flabbergasted to hear his client's attorney direct a question to him that could have come in a well-conceived cross-examination by the opposition. While the author stumbled to answer he glanced around the courtroom. The judge sat stoically, but several jury members were snickering and the opposing attorneys were grinning like Cheshire cats. To gain further time to concoct an answer the author asked that the question be read by the court reporter. Hearing it for the second time merely compounded the agony. This time the judge was smiling. The author answered the question as objectively as he could, but the episode did not win any points for him or his attorney.

The outcome of trials involving environmental-type disasters and failures often boils down to which side's experts are the most believable. Once technically and psychologically prepared, the objective of the expert witness is to impart his or her findings to the judge or jury in a straightforward, easily understood manner. The expert should:

- Be natural. Putting on an act to seem more knowledgeable or sincere is usually transparent and counterproductive. Also, an air of seriousness and professionalism can sometimes be interrupted with an injection of humor if it is part of one's personality.
- Dress conservatively, as if going to the bank to negotiate a loan.
- Talk slowly, distinctly, and into the microphone. Look at the judge or jury when talking, not at the interrogator.
- Remain calm and composed during the cross-examination, even if seething inside.
- Be prepared for attacks or personal discrediting such as accusations of bias, lack of proper qualifications, and exorbitant fees. If an expert has represented the same defendant many times, he or she can expect lengthy and barbed questioning regarding the coziness of the relationship. Conversely, an expert who has worked mainly for plaintiffs can expect insinuations of being a hired gun. For example, at a recent trial

both of the authors were accused by the plaintiff's attorney of being government lackeys. To prove his point to the jury he read from one of our professional resumes, which includes a long list of government clients. Of course he neglected to mention all the other clients, such as developers, contractors, architects, and private corporations. Our attorney corrected this impression in redirect examination.

Attorneys also like to dwell on fees—not their own, just the opposing expert's. It is a no-win situation. If an expert has not spent much time on an investigation, how could the conclusions and opinions be well-founded? If the fees seem too high, this tends to prove the expert was padding the assignment.

In addition there is usually an attempt to find contradictory statements, theories, or opinions based on previous trial transcripts or published papers and books the expert may have authored. In a trial involving big sums of money, an attorney may assign a host of aides in the hope of finding one or two contradictions or inconsistencies; the more trials and publications an expert has under his or her belt, the greater the vulnerability.

Quoting the expert out of context also is common. One of the authors once wrote a published paper on the subject of expert witnessing. In it he made the statement, "You will be working very closely with your client/attorney and you owe that person a high degree of loyalty to do the best job you can." Later, on the same page, the following statement was made: "Although duty and loyalty to your attorney/client are important, they are not paramount. Duty to scientific honesty and personal integrity come first." The opposing attorney chose to read only the first sentence and attempted to construe it as recommending advocacy.

An expert can expect to have to handle a variety of similar challenges. By remaining cool and depending on one's attorney to properly respond, these forays by the opposition can sometimes be turned to advantage owing to their negative effect on the judge and jury. As in any fiercely competitive contest, it is not wise to enter it either underprepared or overconfident.

The principal actors in the initial Big Rock Mesa trial were the sole homeowner plaintiff, his attorney and staff, his insurance company lawyer, the attorneys representing the three defendants (Los Angeles County, Los Angeles County Waterworks District No. 29, and the State of California Department of Transportation [CalTrans]), and the cast of expert witnesses. Prior to this trial nearly 300 other homeowners had initiated lawsuits for property damages and emotional distress against the same defendants. Most of them had also filed claims for damage with their insurance carriers. In nearly all instances the insurance companies had denied payment based on the fact that most homeowners' policies do not cover damage due to land movement.

The attorneys for the waterworks district jumped into the insurance controversy by countersuing all 300 plaintiffs. This action appeared to have a twofold purpose: First, it was probably a sincere attempt to redirect at least part of the blame for the landslides on the individual homeowners and their geologic and geotechnical consultants, that is, their failure to recognize or acknowledge the risks of landslide and their failure to operate and maintain the hydraugers. Second, it appeared to be a legal ploy to drag the homeowners' insurance carriers squarely into the fracas. Instead of falling back on their exclusionary clauses to divorce themselves from the lawsuit, the insurance companies were in the position of having to pay for the defense of their clients—a costly proposition that would help induce an out-of-court settlement. For reasons that we cannot explain, the other two defendants did not join in the countersuit against the homeowners.

It took about 2.5 yr from the time the first lawsuit was filed until the trial began, which is a fairly short interval considering that many civil cases in Los Angeles County take the full 5 yr prescribed by law as maximum. There are simply not enough courtroom judges and supporting personnel to keep up with the caseload. If both parties agree and have the financial resources, time to trial can be reduced by hiring a "rent-a-judge" to hear the case. Many such retired judges are as busy as they were when they were on the bench and are receiving far greater remuneration.

The Big Rock Mesa trial took place in the Los Angeles County Superior Court. By mutual consent the trial was by judge only. It appeared that both sides were willing to let their cases ride on technical and legal merit. Also, it might have been difficult for the plaintiff's attorneys to dredge up jury sympathy for wealthy homeowners.

Total trial time was about 4 months, shorter than most anticipated. The plaintiff's attorneys hammered away at their expected principal themes, namely:

1. That the county knew or should have known the risks of residential development without sewers in the early 1960s and should not have approved the subdivision with septic systems
2. That the county became legally responsible for operating the hydraugers when the homeowners abandoned the project
3. That the county approved an inadequate storm drain design, which allowed too much runoff to infiltrate to the groundwater regime
4. That the county approved all the subdivision plans, including those for streets and storm drains, and then accepted dedication of these improvements for maintenance; therefore, the whole development was, in effect, a county project
5. That water system leaks constituted a substantial cause of infiltration to groundwater
6. That the waterworks district, because of its groundwater extraction

program, never relinquished responsibility for the water through its entire cycle

7. That CalTrans' original grading of the bluff for a slight PCH realignment caused a progressive series of slope failures and landslides that ultimately led to the subject slide

The outcome of the trial seemed to hinge more on the legal issues than the technical. There was no argument that the proximate cause of the landslide or slides was the rapid rise in the water table after development began. Substantial differences of opinion arose as to the geometric configuration of the slide blocks and how the slides started and progressed. However, with the exception of the effect of the bluff failures, these differences did not appear pertinent to the decision. That is not to say that all these landslide studies went for naught; they were of tremendous value in the planning of mitigation measures.

The plaintiff's lawyers seemed to waver on the relative impact of rainfall infiltration versus septic system infiltration. In the early stages of the depositions they appeared to downplay the effect of septic systems and exaggerated the volume of rainfall infiltration. As the trial approached, their tactics seemed to change toward acceptance of a higher volume of septic system infiltration. It was our impression that reversal was due to an emphasis on the culpability of the county in approving the use of septic systems.

As previously mentioned, the plaintiff's expert was not very convincing in his testimony in accentuating the magnitude and importance of water-line leaks. Also unconvincing were arguments by the plaintiff's lawyers that the waterworks district owned and controlled the water that seeped from the septic systems. As a consequence the judge approved the district's attorneys' motion for summary judgment and dismissed the waterworks district from the case. This was done before any district experts had testified. Our role was then reduced to occasional consultation and waiting for the multiple plaintiff trial.

The plaintiff's attorneys were not able to convince the judge that there was a causal link between CalTrans' grading activities along the PCH and the damage to the plaintiff's property. Therefore, a motion for summary judgment by the state's attorney was granted, removing CalTrans from the case.

A similar motion for summary judgment by the county was denied, and the trial continued. The plaintiff's lawyers seemed unable to prove that the county's storm drainage systems had a net detrimental effect on runoff infiltration. At this juncture the doctrines and precedent regarding the county's legal responsibilities became the paramount issues.

Several weeks after the end of the trial, the judge rendered his verdict generally in favor of the plaintiff and against the county. The following excerpts are quoted from the judge's Statement of Decision (actual names have been excluded):

- The Big Rock Mesa evidenced movement in 1978 or earlier. The Big Rock Mesa landslide substantially contributed to the movement of earth under the Plaintiffs' property which caused the damage to the property.
- The County originally contemplated that sewers would be installed for the Big Rock area. The County later allowed development with seepage pits and horizontal drains. The seepage pits and horizontal drains were a substitute for sewers in the Big Rock Mesa area.
- The rise in the groundwater level since the early 1960's was a substantial contributing cause of the Big Rock Mesa landslide, and effluent from the private sewage disposal facilities substantially contributed to that rise.
- The County substantially participated in the planning and approved the development of Big Rock Mesa utilizing seepage pits and horizontal drains. That development contributed to the rise in groundwater which was a substantial cause of damage to plaintiffs' home.
- The horizontal drains required by the County as a condition of issuing the permit for Tracts 26263 and 27463 are public works. The minutes of the Los Angeles County Board of Supervisors are an express acceptance of the horizontal drains to the County as public works.
- The County of Los Angeles approved the drainage system for Big Rock Mesa, but now contends that it did not accept the horizontal drains so they are not a public work. However, the horizontal drains were impliedly, if not expressly, accepted.
- Any sloughing or surficial slides along the bluff did not contribute to the activation of the ancient slide or to the damage to plaintiffs' property.
- The County owns and maintains a number of public roads and associated surface drainage structures on the Big Rock Mesa. They carried water off the Mesa that would otherwise have infiltrated into the subsurface. There is no causal connection between those public improvements and plaintiffs' damage.
- The concept of inverse condemnation stems from the Constitutional mandate that private property shall not be taken or damaged for public use without just compensation. The law as developed by the courts of this state requires compensation to the property owner whose property is damaged by operation of a public work as deliberately designed and constructed, whether or not that damage is foreseeable.
- Our Supreme Court in *Holtz v. Superior Court* (1970) 3 C 3d 296, 311 adopted the analysis by Professor Van Alstyne that ". . . the governmental entity with its superior resources is in a better position to evaluate the nature and extent of the risks of public improvement than are potentially affected property owners, and ordinarily is the more capable locus of responsibility for striking the best bargain between efficiency and cost (including inverse liability costs) in the planning of such improvements."
- The court in *Frustuck v. City of Fairfax* (1963) 212 CA 2d 345 held the City liable in inverse condemnation where it approved maps and plans

which included drainage systems. That conduct constituted a substantial participation incident to serving a public purpose. Incidental to the serving of a public purpose, the County approved the drainage system for the Big Rock Mesa.

- The County was in the best position to assess the risk of residential development of the Mesa with a system of seepage pits and horizontal drains and, under the philosophy articulated in *Holtz,* it must now bear the loss when damage occurs.
- The failure of the Malibu Mutual Drainage Company to carry out dewatering efforts does not relieve the County of liability to these plaintiffs, who are not members of Malibu Mutual Drainage Company.
- The County is liable to the plaintiffs for the damage to their property caused by the Big Rock Mesa landslide and is liable to plaintiffs' insurance carrier for its subrogated interest.
- Plaintiffs rely on two theories to charge Waterworks with liability. First, plaintiffs contend that Waterworks and the County acted in concert in constructing a system for importation and removal of water from the Mesa, and second that leaks in the system were a substantial contributing cause of the damage to plaintiffs' property. Plaintiffs have been unable to sustain their burden of proof on either theory.
- The court rejects the concept that, because the same individuals worked in different capacities for the County and Waterworks in the planning and construction of the system, the project became a joint enterprise.
- Government Code Section 895.2 does not impose liability on Waterworks. Any agreement between it and the County was for services of individuals in the planning and construction of the water delivery system. The construction and planning of that system caused no damage to plaintiffs. There was no agreement for the joint operation or maintenance of the system, and Waterworks had no control over removal of water from the Mesa.
- A review and weighing of the evidence on leakage has convinced the court that the plaintiffs have failed to sustain their burden of proving by a preponderance of the evidence that leaks substantially contributed to the damage they sustained. In addition, there has been no evidence that excludes the probability that other causes alone produced the damage.
- Defendant Los Angeles County Waterworks District No. 29 is not liable to plaintiffs or plaintiff-in-intervention [insurance company] on any theory. Waterworks is entitled to judgment in its favor against plaintiffs and plaintiff-in-intervention.
- The Court is convinced, after a review of the evidence, plaintiffs have failed to sustain their burden that the cuts made on the sea cliff for the construction and widening of Pacific Coast Highway were a substantial cause of damage to their property.
- It has been summarily adjudicated that the plans for the highway were

reasonable and were approved by the State Highway Engineer, and that removal of debris is an exercise of police power. In addition, there is no evidence that removal of debris contributed to the activation of the primary slide. It also has been summarily adjudicated that the four horizontal drains were not owned or controlled by the state and that there was no obligation on the state to maintain them.

- Regardless of whether the toe of the slide is at Pacific Coast Highway or the surf line or offshore, or whether there is one slide or a primary slide accompanied by secondary slides, and there are opinions to support any combination of theories, the Court is not convinced by a preponderance of the evidence that any activity of the state in constructing or maintaining Pacific Coast Highway substantially contributed to the damage to the plaintiffs' property.

- The cuts made by the State may have contributed to sloughing and surficial slides along the bluff, but that did not contribute to the activation of the ancient slide or to the damage to the plaintiffs' property.

- Defendant State of California Department of Transportation is not liable to plaintiffs or plaintiff-in-intervention on any theory. State is entitled to judgment in its favor against plaintiffs and plaintiff-in-intervention.

- Plaintiffs did not pursue separate causes of action for nuisance and dangerous condition of public property but they contend that those theories of liability would support a finding of inverse condemnation. The court finds that there is no evidence to support recovery under either theory.

- Plaintiffs have failed to establish a breach of mandatory duty by the County. It was summarily adjudicated that Section 308(b)(2) of the County's Building Code imposes no mandatory duty at all. Section 308(b)(1) imposes no duty on the County, mandatory or otherwise, in the absence of a finding that "property outside the site of the proposed work could be damaged by activation or acceleration of a geologically hazardous condition and such activation or acceleration could be attributed to the proposed work on or change in use of, the site for which the permit is requested." The County Engineer knew before 1963 of the existence of a prehistoric landslide, but there is no evidence that could be interpreted that the County or any of the experts who investigated the Mesa thought at any relevent time that a geologic hazard within the meaning of Building Code Section 308(b) existed.

- The reactivation of the ancient landslide was considered a remote possibility. The developer's geotechnical firm calculated a safety factor of 2.9 with low groundwater and 2.1 with high groundwater, both well above the standard of 1.5 required by the County, and undisputably reasonable. Most of the experts who testified in this case had all of the information available to the County and, until a slide occurred, were of the opinion that the reactivation of the ancient slide was a remote possibility.

- Neither the County nor the Flood Control District owned or controlled any wells on the Mesa.
- At all material times, the only public improvements owned, controlled, maintained, or operated at Big Rock by the Flood Control District consisted of four storm drains. The drains properly performed the functions for which they were designed and constructed, and each of them resulted in less water reaching the groundwater table than would have occurred without their installation. Accordingly, no Flood Control District public improvements were a cause of plaintiffs' damage.
- The Flood Control District played no part in planning or approving private construction at Big Rock, nor was it involved in the decision to allow development without sewers. Accordingly, the Flood Control District is not liable to plaintiffs or plaintiff-in-intervention and is entitled to judgment against plaintiffs and plaintiff-in-intervention.
- Plaintiffs are not entitled to recovery for emotional distress under the law of inverse condemnation. Moreover, such relief is unavailable in conjunction with property damage in the absence of a special pre-existing relationship or an intentional tort, neither of which is present here. Nor is emotional distress recovery available to plaintiffs under a failure to warn theory. The County had no duty to warn plaintiffs of the Big Rock Mesa landslide, nor was that danger foreseeable by the County.

Attorneys for the county immediately began the appeal process. It was nearly 2.5 yr from trial court decision to the state appellate court's decision. The appellate court generally affirmed those portions of the trial court's decisions covering the defendants (CalTrans, the waterworks district, and the flood control district) *but reversed that part of the decision against the county.*

The appellate court document is extremely lengthy and loaded with many legal precedents not germane to this book. Although we do not consider it appropriate to reprint the entire decision, we do find the comment at the end of the decision to be representative of the philosophy of the appellate court:

> Though unnecessary to our decision, some comments seem appropriate. The increasing social tendency to blame the government for events which are largely the result of abandonment of individual responsibility is not to be encouraged by so-called advances in the law. Here private developers sought to reap the profits from building on hillside land without bearing the expense of a sewer system; the developer convinced the County this would be safe, especially because a private dewatering program would be maintained. The residents who eventually occupied the area, however, chose to abandon their private dewatering efforts, apparently in hopes a public agency would take over and save them some modest expense. That the eventual catastrophe should be followed by an effort to shift the loss to the public is unsurprising but deplorable.
>
> The notion that the County is liable in inverse condemnation for permitting a

development it should have banned is ironic, since if the County had refused to permit the development, or required installation of a sewer system or other conditions which the developer considered unreasonably restrictive or costly, the County would undoubtedly have been sued in inverse condemnation for a taking of the developer's land. (See *Frisco Land & Mining Co. v. State of California* (1977) 74 Cal. App. 3d 736, cert. denied, (1978) 436 U.S. 918; *MacDonald, Sommer & Frates v. County of Yolo* (196) 477 U.S. 340.)

The inverse condemnation claim has an additional intrinsic absurdity: the County is said to be liable to plaintiffs for its having permitted their neighbors' homes to be built with the same sewage disposal devices as their own.

As aptly stated in *Yox v. City of Whittier* (1986) 183 Cal. App. 3d 347,355, "It would be unfair and contrary to the relevant constitutional provisions to inflict upon the community as a whole liability for a loss resulting from a strictly private improvement." The government is not our insurer of last resort.

Disposition

The judgment in favor of the plaintiffs against the County of Los Angeles is reversed, with directions to enter judgment for the County of Los Angeles. The judgment in favor of Los Angeles County Waterworks No. 29 is affirmed. The order striking the cost memoranda of the Los Angeles County Food Control District, Los Angeles County Waterworks No. 29, and the State of California, are reversed, with directions to award costs to those defendants and to the County of Los Angeles. The County of Los Angeles, Los Angeles County Flood Control District, Los Angeles County Waterworks No. 29, and the State of California are to recover their costs in these appellate proceedings.

Between the time the appeal was filed and the appellate court decision rendered, two other trials took place. Both cases involved plaintiffs who were over 70 yr of age. California law requires that litigants over 70 yr old be given priority in the court docket. CalTrans, the lone defendant in the first trial, was found liable by a jury, principally on the argument that the instability of the PCH bluff was an proximate cause of the landslide. The second case, also directed at CalTrans, was settled during the trial.

While the attorneys for both sides were girding themselves for the multiple plaintiff trial, the appellate court decision came in, apparently creating the atmosphere for a settlement. When the public agencies and insurance companies offered a combined total of $97 million (averaging about $200,000 per plaintiff after attorneys' fees), the homeowners accepted. The settlement represented only about 20–25% of their claims, but, in view of the appellate court decision in the trial case, to go for the quick money without risk seemed the prudent thing to do.

As with many out-of-court settlements, none of the parties appeared particularly happy. Meanwhile the homeowners faced a cost of about $100,000 each to finance a special assessment district for the purpose of continuing to dewater Big Rock Mesa. Controversy regarding the assessment district is ongoing.

THE POST MORTEM

The Big Rock Mesa landslide may have spawned more controversy in its aftermath than any forensic undertaking in our experience. The following topics for debate are among the many that stemmed from this lawsuit.

Failure to Recognize or Acknowledge Existence of a Landslide

The failure of at least 10 consulting geologists, plus several employed by the county, to recognize Big Rock Mesa as a landslide as distinguished from a marine terrace was the root cause of this disaster. It is difficult to believe that this number of experienced and supposedly well-qualified geologists could not make this distinction correctly.

Consulting geologists hired by developers have a tendency to conduct a cursory investigation of a hillside terrace such as Big Rock Mesa rather than fully evaluating the possibility that the terrace may have developed as the result of landsliding. This bias (or fraud) stems from the shortsighted notion that the identification of a landslide on a site will increase the projected cost of the development to the extent that it might become unprofitable. In fact, the identification of landslides and other adverse site conditions is a fundamental step in the determination of development costs. It is not the role of the geoscience consultant to worry about the financial viability of a project during predevelopment site investigations. Failure to identify and remedy adverse site conditions ultimately leads to greatly increased cost in the form of mitigation, legal fees, and possible punitive damages levied against the developer and the subcontractors after a geologic disaster occurs on the site. Case history studies conducted by one of the authors show that making the correct analysis during the preliminary planning of a project can result in a cost : benefit ratio of from 1 : 100 to 1 : 1000. Big Rock Mesa seems to have been considered a wave-cut marine terrace primarily because that was the interpretation that was most desired of those who were involved in its residential development in the 1960s.

As for the county geologist's failure to recognize the landslide, the following explanations (not excuses) might be advanced:

1. It was not (and still is not) county policy to provide its own independent geologic investigation and analysis for a project. The county's role is basically to review for conformance to grading requirements and code regulations. While its employees can and sometimes do question items of procedure and completeness of scope in a geologic report, they seldom question the conclusion and recommendations. This policy should be changed with a view toward expanding the review and approval role of the public bodies having jurisdiction.

2. While the county geology staff had no reason to curry favor with the developers, there is often pressure, albeit subtle, from elected officials to approve questionable projects.

3. The competence level of geologic practice is still being established as "standard" rather than "good." Existence of a landslide hazard could have been recognized and abated had good scientific procedures been used. For example:

- The county did not and does not require the use of aerial photographs.
- The county did not and does not require that adjacent/nearby properties, which could provide clues, be examined for geologic data. It suggests that only the property in question be a part of the analysis.
- Most consultants are advised by their attorneys to look at, review, and analyze *only* the property they have been employed to investigate and never to comment on property owned by others. This is because they may be sued by the owners of those other properties. With this procedure the true geologic environs and hazards are seldom ever recognized.

Establishing Proper Design Priorities

A fundamentally flawed design policy that we have seen repeated again and again is the sequence of consultation a land developer often uses when building on hillside property: (*a*) architect/land planner; (*b*) civil engineer; and (*c*) geologist and geotechnical engineer.

Typically the architect/land planner first creates a scheme of development that accentuates aesthetic and market considerations. The civil engineer then comes aboard to refine the scheme and to design the earthwork, storm drainage, streets, and utilities to conform to applicable codes and regulations. In the process the more creative civil engineers find ways to maximize the area of usable land and the number of lots, the bottom line for developer profitability. The developer is then able to estimate construction costs and prepare a financial pro forma. At this point the developer is often belatedly reminded (or has been aware but reluctant to act) that geologic and geotechnical engineering reports are mandatory in most jurisdictions. The fact that the scheme has already been cast in stone puts the geologist and geotechnical engineer in an awkward position. Their investigations may reveal:

- That certain predevelopment corrective work should be done such as removal of soil material not suitable for fill, removal of slide debris, excavation and recompaction of alluvium in canyons, and installation of subdrains
- That some areas should be only partly developed or not developed at all because of the existence of faults, landslides, etc.
- That further investigation of potential landslides is warranted

- That proposed cut-and-fill slopes may have to be flattened, thereby lessening the yield of usable land
- That fills need to be buttressed

Introducing any of these items so late in the planning process could result in developer apoplexy. The architect/planner and civil engineer will be unhappy, too, particularly if their fees are based on anything other than time and materials. Faced with costly and disruptive changes to the scheme, the developer, designer, and sometimes local government officials may apply great pressure on the geologist and geotechnical engineer to reject or at least drastically modify their findings, opinions, and recommendations. (After all, many of these recommendations are based on scientific concepts and are not mathematically precise.) If the geologist and geotechnical engineer cannot be budged, it is a good bet that the developer will be able to find others who will have a more cooperative philosophy.

Many such unhappy and hazard-fraught situations could be avoided by reversing the order of consultant employment. If the geologist and geotechnical engineer were hired to work with the architect/planner, problem areas could be identified in advance of development concepts. The civil engineer could then prepare tentative grading, road, and utility plans based on the geologic, soil, and topographic parameters as well as on the concept planning. From this point forward, all parties could work in harmony to prepare final plans that not only would reflect maximum land yield and architectural and marketing considerations but would provide maximum safety and minimum exposure to future liability. Our experience indicates this methodology has worked very successfully. Unfortunately its use requires sophisticated developers possessing long-term outlooks, and they are in the minority.

Wisdom of Using "Performance" Specifications

Specifications or conditional permits for construction fall into two general categories:

1. "Prescriptive" specifications, which spell out what is to be done and exactly how to do it
2. "Performance" specifications, which identify the results to be obtained with little or no reference to methodology

The first approach is generally used whenever it has been firmly established that a particular method or procedure will yield the desired result. When the outcome of a precise method is uncertain, it may be far more effective merely to specify the desired result and to allow the contractor or developer to use any procedure that does not adversely affect other aspects of the project to achieve those objectives. The City of Los Angeles has utilized many performance specifications in its code with good results.

In retrospect the county should have insisted on a performance specification for a groundwater extraction program when conditioning the approval of the original Big Rock Mesa subdivisions. Specifying four hydraugers 700-ft (215 m) long was, apparently, an arbitrary recommendation of the consulting geologist. To our knowledge no quantitative estimates were ever made regarding the volume of water they would drain or their effect on the water-table levels. Neither was there a groundwater budget prepared nor even an estimate of infiltration from rainfall, septic systems, and irrigation. A performance specification, on the other hand, would have merely required the developer and future homeowners' association to install, operate, and regularly monitor whatever was necessary in the way of wells and hydraugers to maintain specific maximum groundwater elevations at particular locations. Further, the county should have stipulated that in the event of nonperformance by the developer or the homeowners' association the county would take over the operation and maintenance of the facilities and charge the homeowners for the service.

If this procedure had been followed the entire disaster could have been averted. It is interesting to speculate what effect such a specification in the conditional approval would have had on the developer. Undoubtedly he would have screamed that he was being burdened with an unfair, unnecessary, and unpredictably costly requirement that would price his finished lots out of the market. If his protestations were not heeded at the staff and administrative levels, he probably would have carried his fight to the County Board of Supervisors. And if that appeal had been unsuccessful, he may have attempted to sue the county. The outcome of such a lawsuit would have hinged on whether the specified maximum water-table levels could have been justified in terms of the potential hazard. Based on the nonrecognition of a deep-seated landslide at that time by the developer's consultants, it might have been difficult to establish maximum safe groundwater levels. But shouldn't the county opt for safety over an increment of developer profit? Besides, in the annals of hillside development in popular areas like Malibu, such necessary added costs are easily passed on to the buyers. And isn't this the way it should work?

Even if a lawsuit had resulted, it would have been far better to settle the issue before development, when the litigation stakes were miniscule compared with those in 1983. Besides, well-prepared defenses against such lawsuits have been successful.

The key question, of course, is whether the homeowners would have abandoned this more expensive method of controlling groundwater levels just as they did the hydrauger program. If so, the results might have been the same. However, the county might have been in a better legal position since the homeowners would have abandoned a more effective system than the four hydraugers (which may not have prevented the landslide even if they had been operated and maintained properly).

Public versus Private Works

That the homeowners were able to walk away from what appeared to be a reasonable obligation to maintain the hydraugers raises serious doubts about the wisdom of making private organizations responsible for vital quasi-public works systems that require diligent operation in perpetuity. Instead, the county should have created an assessment district among the property owners on Big Rock Mesa to pay for the operation and provided for maintenance of the system through a property tax assessment. However, the actual responsibility for the system should have rested with the county.

With the advent and rapid growth of planned private communities, condominium developments, etc., government has had to evaluate what should remain public works (or public utilities) and what can be transferred to the property owners. The criteria should be predicated on public health and safety considerations. The operation of the following public works could adversely affect health and safety:

- Sanitary sewer mains
- Sewage pumping stations
- Sewage treatment or reclamation facilities
- Water supply and distribution facilities
- Electrical and gas utility facilities
- Major storm drainage facilities
- *Dewatering facilities*
- Major erosion control facilities
- Primary and secondary streets, including traffic signals and lighting

Property owners associations might own and be responsible for operating and maintaining:

- Local streets
- On-site sewers and storm drains
- Parks and landscaping
- Minor erosion control facilities
- Irrigation systems

There is no question in our minds that the responsibility for operation and maintenance of a dewatering system that is intended to help stabilize land should lie with a public agency. Or, at least, the system should revert to a public agency when the private sector fails to act prudently.

The Sewer Dilemma

As Malibu grew more densely populated its residents became increasingly aware of the health hazard posed by the use of individual septic systems. It appears the health issue has never really been a subject for debate. However, most Malibu residents, when faced with the dilemma at the polls, have voted for what they felt was the lesser of two evils. They perceived the health hazard to be less onerous than a growth-inducing, costly, and environmentally questionable sewer system complete with treatment plant and ocean outfall pipe. Many, with a modicum of justification, have pointed to the example of the huge City of Los Angeles Hyperion Wastewater Treatment Plant and how its frequent malfunctions have abetted the pollution of Santa Monica Bay.

What has been mysteriously absent or downplayed in the 25-yr debate over sewers is the role of hillside septic systems in creating imbalances in land stability. Many Malibu landslides have occurred in undeveloped areas triggered only by excessive rainfall infiltration. Faced with this natural delicate balance, introduction of septic systems with their substantial and steady effluent input to the groundwater regime have been and will continue to be a major contributing factor to instability. Somehow it does not make sense to permit a huge release of subsurface effluent and then require its extraction before it can do any harm; a procedure not only inefficient and expensive but subject to myriad human and mechanical errors. Also, in the near future extracted water will have to be treated before it can be discharged into the ground, into streams, or into the ocean. Coliform bacteria content of water pumped from the wells at Big Rock Mesa as part of the mitigation program continues to increase.

In Malibu, as well as in many other naturally unstable or marginally stable hillside areas, sewer systems are an absolute necessity. There are only two alternatives: One is to prohibit development in these areas or, at least, drastically restrict development. Such extreme action would, of course, invite lawsuits. The other alternative is not attractive either. It would involve purchasing or condemning such land and converting it to parks or permanent open space, which would be a gigantic financial undertaking also fraught with legal obstacles.

Finding Other Means of Reducing Water Infiltration

It is clearly evident from the Big Rock Mesa experience that the planning and design of hillside development should incorporate every feasible method of reducing water infiltration to the groundwater regime. Some of these methods include:

- Mandatory use of roof drains that discharge rainfall to streets, underground storm drains, or lined open drainage courses. The City of Los Angeles has had such a requirement in its building code since 1963. The code item has probably added less than 1% to the cost of a typical hillside residence but has been tremendously cost-effective in helping to prevent both erosion and excessive subsurface infiltration. Other jurisdictions have been slow to follow suit, not only because of developer resistance but because many hillside residents believe roof drains and downspouts are not in keeping with a rural, rustic setting. In hillside areas, safety and antilitigation measures must receive the highest priority. Roof drains were not a requirement in Big Rock Mesa (nor are they mandatory to this date anywhere in the unincorporated hillside areas of Los Angeles County). Only a few of the Big Rock Mesa homeowners had installed them, probably to eliminate nuisance ponding adjacent to their homes.

- Grading codes that dictate that all landscaped areas have sufficient surface gradient to prevent ponding of water
- Generous use of underground storm drains and lined channels. Again, there is often serious conflict with aesthetic considerations, but safety should prevail.
- Use of ground moisture control sensors to prevent overirrigation of automatically sprinkled landscaped areas

When combined with the installation of sanitary sewers, these measures would at least ensure that development of a geologically sensitive hillside area would not create additional instability over what existed in its undeveloped state.

To Remove or Not to Remove Roadway Debris—A Catch-22

A nagging problem for public agencies charged with the maintenance of roads in landslide-prone hillside areas is that when a landslide occurs that deposits debris on a roadway, the agency is obligated to clear the debris to permit public use of the road. This is usually a continuing maintenance function since most landslides continue to move. The catch is that removal of the earth may remove a small measure of lateral support and, in a very minor and usually insignificant way, contribute to further sliding. This fact has been meticulously exploited by plaintiffs' attorneys in filing lawsuits against public agencies.

Faced with this dilemma, public agencies could help themselves by adopting certain policies:

1. When a landslide occurs, investigate immediately to see if there are alternative routes available to serve all the affected properties, even if rerouting incurs additional driving time and inconvenience. If rerouting is feasible, the road should be barricaded on each side of the slide and no debris should be removed until the landslide has been judged stabilized.

2. Utilize the professional services of a competent engineering geologist and geotechnical engineer to evaluate the consequences of removing the slide

debris from the roadway. If there is any question that removing the debris could accelerate the slide and either hasten or worsen slide damage, the debris should be left in place, even if it necessitates special traffic controls or construction of new access roads. If the professionals' opinions are that removal of the debris would have an insignificant impact on the slide, the agency would at least have an unbiased professional report to use in the event of future litigation.

3. There should be detailed documentation of the debris removal consisting of photographs taken before and during removal and a record of the dates and volumes of removal. Again, if a lawsuit ensues, this data could prove extremely useful.

In the case of the Big Rock Mesa landslide, CalTrans had no choice but to remove the debris from the PCH as quickly and expeditiously as possible. The PCH is a major state thoroughfare for which there is no alternate route and which, for topographic reasons, does not lend itself to temporary detours.

The deep-seated nature of the Big Rock Mesa landslide, the slide plane being about 230 ft (69 m) below the PCH, makes it evident that the slide was not affected by the removal of the debris. However, this fact does not mean its removal did not have some effect on the surficial sloughing of the bluff slope, which was probably a major confusion factor in the lawsuit settlement negotiated by the state's attorneys.

Water-Line Ruptures

The fact that the Big Rock Mesa landslide occurred slowly over several months allowed the waterworks district sufficient time to replace its underground lines with above-ground flexible-jointed pipes. This action not only reduced the number of leaks and ruptures but provided for ease of inspection and repair.

In the event of fast-moving landslides there is inevitably at least one pipe rupture that permits large volumes of water to infiltrate the slide mass. If litigation ensues there is always the "chicken and egg" question of whether the pipe rupture occurred first and helped to precipitate the landslide or whether the pipe rupture was caused by the landslide. If the sequence of the two events cannot be established on the basis of direct evidence, such as an eyewitness account of unusual surface moisture or the timing of a sudden drop in water pressure, the forensic investigation should include such items as:

1. History of water-line breaks or failure in the area
2. Pipe material and age
3. Water-pressure variations

4. Depth of trench, bedding conditions, and basic fill materials
5. Analysis of volume of water that escaped the rupture
6. Chemical analysis of the water compared with an analysis of the groundwater
7. Geologic conditions near point of rupture relative to ease of underground migration
8. Existence of previous landslides, both ancient and recent
9. Detailed engineering geologic study including borings and testing to determine depth of slide surfaces and underground water levels
10. Geotechnical calculations of slope stability under various moisture conditions and geometric projections.

In most cases we have found that the slide caused the pipe rupture. This does not imply that the reverse cannot be true, particularly if the main is large, the rupture is severe, and there is a considerable time lapse before the flow is stopped.

Land-Planning Philosophy

Like most landslides the Big Rock Mesa slide proved a tragedy to the homeowners, a time-consuming headache to public officials, a financial burden to the county taxpayers, and a financial bonanza to the attorneys. It seems to us that few people in local government are aware (or willing to admit) that Big Rock Mesa represents the tip of the iceberg as far as potential landslide losses in Malibu, and other hillside areas with similar geologic characteristics, are concerned.

In our opinion two firm actions must be taken, not only in Malibu but in other landslide-prone regions where there is great development pressure, to provide orderly and safe development. The first recommendation is stopgap in nature. The second represents a long-term solution.

1. There must be a change in building department philosophy, policy, and procedure whereby building permits for individual parcels are not necessarily issued on the basis of a single favorable report by the developer's geologist. Landslides do not recognize property boundaries, and geologists are seldom paid by an individual lot owner to investigate an entire zone of influence.

It is our experience that geologic reports that focus on individual isolated lots or parcels, except for those by a few competent and honest geologists, are practically worthless. In fact they are often harmful because they can create a false sense of security on the part of both the developer and the owners of adjacent property as well as local government officials. In addition to the Big Rock Mesa homeowner who was the plaintiff in the original lawsuit, two other personally experienced flagrant examples of building permits being issued for individual parcels based on a favorable geologic report come to mind:

- We were engaged as expert witnesses in a series of lawsuits involving a landslide located in what is known as the Rambla Pacifico area of Malibu. The boundaries of this 1983 slide were very apparent in aerial photographs dating back to 1927, were physically evident as early as 1978, and were unmistakable by 1980. In spite of these facts a favorable geologic report in 1980 for an individual property owner, located at the headscarp of the slide, resulted in a building permit being issued. Construction of the house was completed in 1982. It was occupied for only a few months before being completely destroyed by a massive landslide. In his report the geologist acknowledged the existence of the potential slide but inferred that the house would be safe because its footings would be founded on an island of solid bedrock, not part of the slide mass. The homeowner filed suit against the geologist, but since the geologist's pockets were "shallow" and he carried no insurance, the action was dropped.
- Another outrageous example of tunnel vision geology concerned an individual hillside home in the Santa Monica Mountains. We had been asked by the owner to inspect the lot because she felt uneasy about a favorable geologic report that she had obtained when she bought the property. Upon inspection it was immediately obvious that there was a serious landslide threat from a slope on undeveloped land across the street. In reviewing the geologist's report we found no mention of the threatening landslide, undoubtedly because it was outside the subject property boundaries. The only reference in the report to any potential hazard was the observation of gopher holes in the backyard. What made this example of tunnel vision so difficult to understand was that the owner thought she was paying for a professional report that would address her interests and concerns. It was not as if the report had been purchased by the seller of the property.

2. All of Malibu (as well as other comparable regions) should be investigated by state and federal geologists and reviewed by an impartial blue-ribbon panel of geologists and geotechnical engineers employed by local governments. The investigation should include all necessary borings and field and laboratory testing to permit reasonably detailed mapping. The end product would consist of geologic maps delineating known existing and potential landslides and other land hazards. They would be not unlike the USGS landslide maps but would be more comprehensive and detailed, particularly in indicating probable slide geometry and estimated depths of slide planes.

Professional engineering geologists would expand on this data and subsurface borings and analyses. Geotechnical engineers would calculate slope-stability safety factors for all the slides using strength parameters obtained by sampling (in conjunction with the geologists' borings) and testing the weakest materials within the various projected geometric schematics for the landslide. An analysis would also be made of the most feasible mitigation concepts for the various slides, including preliminary cost estimates.

Armed with such information, government officials and land-use planners could intelligently evaluate plans for future development.

- Areas that are the most hazardous and most expensive to mitigate might be purchased and used as open areas and parks or left in their natural state until land values would warrant proper mitigation. Valuation appraisals for these areas could be predicated on the impartial expert report of estimated cost of repair and prevention as well as normal development costs. This would result in more realistic values than when appraisers are not aware of the hazard or the cost to remedy.
- In other areas land-use densities could be at least partly based on the degree of hazard, with clustering of development in the stable areas and avoidance of the hazardous areas.
- Individual lot owners within landslide areas could band together to provide mitigation measures. Local government could assist by creating assessment districts so that costs could be fairly prorated among property owners, spread over a greater period of time, and paid as a property tax assessment. The California Legislature recently passed a law that provides the legal mechanisms for the establishment of such landslide abatement districts. Without a doubt these procedures would slow the pace of development by requiring abatement and mitigation. Initially the requirements would probably result in lawsuits against local government. But these lawsuits would be much smaller in dollar amount, much easier to win, much less expensive to lose, and much cheaper to defend than those brought about as the result of landslide damage after development had occurred.

Our experience convinces us that in the event of landslide damage, regardless of circumstances, plaintiffs' attorneys will always find a way to point the finger at the government with its "deep pockets." It therefore behooves government to prevent hazardous development by any appropriate method.

4   UTAH LANDSLIDE: RURAL SLIDES CAN BE DEVASTATING TOO

THE EVENT

High in the picturesque Wasatch Mountains of north-central Utah there once existed a tiny railroad village named Thistle. Use of the past tense denotes that in 1983, after surviving more than a century of economic perils, Thistle was totally obliterated and temporarily replaced with a huge lake as the result of a massive landslide that dammed the Spanish Fork River.

Thistle's origin dates back to the early 1870s, when it served as a campsite for the construction of the narrow-gage Utah and Pleasant Valley Railway. Nestled in a valley 5000 ft (1500 m) high known as Spanish Fork Canyon, Thistle was ideally located for the construction of permanent railroad maintenance and operation facilities.

In 1890 the Denver & Rio Grande Western Railway Company (D&RGW) purchased the railroad and converted it to standard gage. As D&RGW expanded its operations, Thistle became a vital main-line station between Denver and Ogden. Its population steadily grew, peaking at about 600 in 1917. After World War II, as diesels began to replace steam locomotives, many of the town's facilities became obsolete. By 1983 there were only about 50 residents and 20 homes left in Thistle. Most of the remaining residents were former railroad employees who did not want to leave the peaceful valley and its natural beauty and railroad-town ambience. When disaster struck, it proved fortunate that the population had severely dwindled.

In 1912 several companies operating coal mines in Carbon County formed their own railroad, the Utah Railway Company (URC). The plan was for the railroad to transport coal from the mines to Ogden via Spanish Fork Canyon. In so doing its tracks would generally parallel those of D&RGW. Because D&RGW owned the right-of-way through the narrow pass west of Thistle, it was necessary for URC to tunnel its track through Billies Mountain. After the tunnel had been excavated about 300 ft (90 m), URC officials had second thoughts and decided to negotiate an agreement with D&RGW to share facilities through the valley. Terms of the lease agreement called for D&RGW

to add another track from Provo to the URC junction in Carbon County. One of the tracks between Provo and Thistle would be owned by URC. Seventy years later legal interpretations of this lease agreement would become an important matter of adjudication between the companies.

Figure 4–1 illustrates the physical relationship of the Thistle area to the

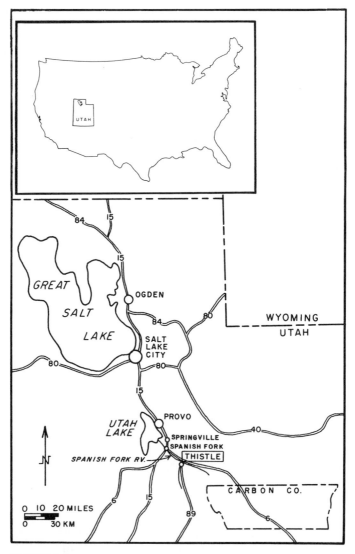

FIGURE 4–1
Vicinity map of the Thistle landslide.

surrounding communities, railroads, highways, and natural watercourses. Spanish Fork River forms at the confluence of Thistle Creek and Soldier Creek, from where it flows northwesterly and empties into Utah Lake, a tributary to the Great Salt Lake. The total area of the watershed contributing runoff to the Spanish Fork River at Thistle is over 500 mi^2 (1300 km^2). Not unlike most western rivers, flow in the Spanish Fork is highly seasonal, varying with the intensity and duration of spring rains, depth of winter snowpack, and timing of spring thaws.

Localized topography existing prior to 1983 is shown in Figure 4–2. Figure 4–3 is a transverse cross section of the canyon at a point several hundred feet downstream from Thistle. This is the location D&RGW denotes as Milepost No. 681.4. The cross section shows the cut slopes on the west side of the canyon created by construction of the railroad and of U.S. Highways 6 and 89.

For many years officials and employees of the D&RGW had apparently noticed ground movement near the tracks at Milepost No. 681.4 as well as at several other locations. Geologists not employed by the D&RGW had identified the area that was to become the Thistle landslide, adjacent to Milepost No. 681.4, as an ancient landslide.

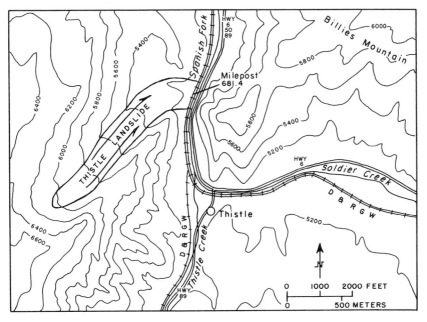

FIGURE 4–2
Site map of the Thistle landslide.

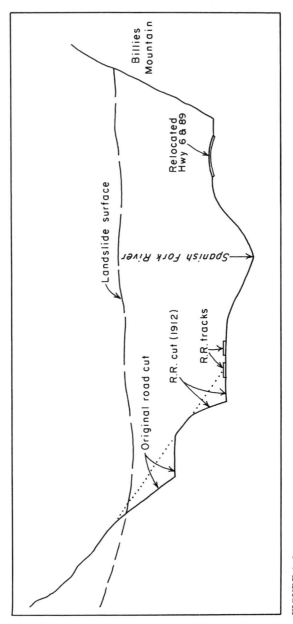

FIGURE 4–3

Schematic cross section at Milepost No. 681.4 looking downstream.

During the railroad's early years there was apparently not enough movement to warrant remedial action. However, sometime in the 1950s a portion of the toe of the frontal lobe of the slide apparently encroached on the west side of the tracks to the extent that the railroad was forced to remove several thousand cubic yards of slide debris. Subsequently, springtime maintenance work was required periodically as the tracks were progressively uplifted and shifted to the east. The incremental vertical and horizontal realignments eventually resulted in the tracks at Milepost No. 681.4 being about 5 ft (1.5 m) higher than original grade and about 3 ft (0.9 m) easterly of the original horizontal alignment. Remedial realignments had to be frequently accomplished to provide for smooth vertical and horizontal transition curves.

In late September 1982 intermittent rainstorms of considerable magnitude hit the area, the result of a large tropical storm originating off Baja California, Mexico. Total rainfall in the Thistle area from September 1982 to April 1983 (including snowfall equivalent) was considerably above normal. In addition, the winter of 1982–1983 was generally mild in temperature, so a greater than usual percentage of the precipitation was rainfall rather than snowfall, or snowfall that melted before the spring thaw. Consequently infiltration to the groundwater regime was exceedingly high. By early April 1983 ground saturation had probably reached record proportions. Rainfall volumes during the remainder of April were also greater than normal, exacerbating the groundwater problem.

In previous springs, particularly those preceded by heavy precipitation, numerous small, relatively shallow slides emerged on the face of the westerly slope near Milepost No. 681.4. They were located in the natural slope as well as on the highway and railroad cut slopes. The 1982–1983 season was no exception. However, the number, surface area, and depth of the slides were of greater magnitude. Also, a scarp appeared near the top of the natural slope that continued to widen and deepen.

As spring approached D&RGW personnel were becoming increasingly concerned about the stability of their cut slope directly above Milepost No. 681.4 as well as surficial failures at the mileposts along their right-of-way. The area was visually monitored on virtually a day-to-day basis. By the first week in April concern had grown to the point where the problem commanded the attention of the top executives of the railroad. A prominent consulting geologist was employed to make a field inspection on April 12. It was too late for mitigation. The next day new noticeable landslide movement began. The tracks were pushed about 3 in. (7.6 cm) out of horizontal alignment. Efforts to realign them were unsuccessful as the motion continued. Heavy earthmoving equipment was ordered to the site. By that evening the highway had heaved and buckled. Highway traffic would continue intermittently for another day until a bulge 11 ft (3.3 m) high would close Highway 89. That same day rail service was discontinued.

As the landslide motion accelerated the chief worry was to keep the streambed at least partly open so as not to impede flow in the Spanish Fork River. By this time it was no longer just a concern for the railroad and highway officials. Blockage of the river would create two major problems; one was immediate and real and the other potential. The immediate concern was that blockage would rapidly back up the higher than normal flow and inundate the structures in Thistle. Since there would be a few days' time to evacuate the 20 homes, this problem, however serious, was not life-threatening. The other problem was that if the river flow could not be kept open and if the rapidly forming natural dam and water it held back built up high enough, the increasing water pressure could cause a breach in the dam. Under this scenario there could be sudden and deadly downstream flash flooding, especially in the communities of Spanish Fork and Springville.

By this time the State of Utah (other than the Utah Department of Transportation) had become involved in the emergency planning process. More heavy earth-moving equipment was rushed into Thistle from all over the state in a frantic attempt to keep the water flowing. At the same time residents of both Thistle and the downstream communities were being warned to prepare for possible evacuation.

The valiant effort to keep the river flowing was doomed to failure from the start. There was simply too much slide debris moving down into the canyon to cope with. On or about April 16 the earth-moving activity was abandoned. Now the water level in the lake, created by the impounded water, rose rapidly. By April 16 the residents of Thistle were evacauted and by the next day their homes were completely inundated.

Once the natural dam became a fait accompli and was still growing, in both breadth and height, priority attention turned to draining the lake before it could threaten the dam. The temporary solution consisted of a battery of pumps connected to flexible conduits laid over the crest of the dam. The permanent drainage scheme was to bypass the dam with a tunnel 2250 ft (675 m) long through Billies Mountain. The tunnel lining was composed of corrugated metal pipe 12 ft (3.6 m) in diameter. The upstream intake consisted of a floating trash rack and a vertical, concrete-lined drainage shaft connected at right angles to the tunnel pipe. Downstream the tunnel connected to another vertical shaft that discharged into the river at a point below the dam.

To control water-flow volumes a concrete bulkhead with two control valves was constructed in the tunnel. The bulkhead and valves would be removed in times of spring runoff to allow unimpeded flow. Before construction was completed the slide had become 1200 ft (360 m) wide and 6800 ft (2040 m) long and the landslide dam had attained monstrous dimensions of 220 ft (67m) high, 900 ft (275 m) wide, and 1500 ft (460 m) thick at the base. The lake had extended to a length of 3 mi (5 km), attained a maximum depth of about 200 ft (61 m), and contained over 64,000 acre ft (3800 ha m) of water.

Considerable debate ensued about whether the lake should be preserved

Winter scene of the massive Thistle, Utah, landslide and the lake created when the slide dammed the Spanish Fork River. The town of Thistle once existed about 50 ft (15 m) below lake-surface level at the upper left portion of the photograph.

for recreational use. There was even talk of creating a hydroelectric generating facility. Utah County and the town of Spanish Fork were very much in favor of keeping the lake. However, the state was concerned about the long-term structural integrity of the dam and exercised its prerogative to drain the lake. The draining process was completed in December 1983.

Buttressed by Billies Mountain, the landslide eventually reached a quasi state of equilibrium. As a result of the slide it was necessary for D&RGW and URC to reroute their rolling stock. For D&RGW it was estimated that the cost of the blockage was about $30,000 per day. Consequently railroad executives and their consultants met in nearly continuous emergency sessions to evolve a plan of action. There were really only two alternatives: to abandon the line and negotiate alternate routes on a permanent basis, or to bypass the dam by tunneling through Billies Mountain. Although construction costs would be unpredictable, it was decided to tunnel.

Once the decision was made to reroute the railroad by tunneling through Billies Mountain, the entire project was designed and built in 81 days—an incredible feat reminiscent of the "can do" Navy Seabees during World War II. Work was performed 7 days a week around the clock in 8-hr shifts. Two major construction contractors were employed. At one time there were 650 workmen on the project. Coordination and cooperation between the two companies under the alert direction of D&RGW officials were incredible. The first train passed through the tunnel on July 4, which provided double cause for a spirited celebration.

Since the tunnel was only large enough to accommodate one track, it was necessary to excavate a second tunnel. There was not quite the same urgency as with the first tunnel, so the second took about 5 mo to complete. Total construction cost of the two tunnels and the 3 mi (5 km) of realigned tracks was estimated at about $40 million.

The speedy reconstruction and realignment of Highways 6 and 89 were no less remarkable than that of the railroad tunnels. Design and construction occurred concurrently, which is sometimes a necessity with a disaster. Total excavation and embankment was in excess of 6 million yd^3 (4.8 million m^3). To meet a target completion date of December 1, 1983, the contractors had to move as much as 50,000 yd^3 (40,000 m^3) per day, an unbelievable volume considering the rugged terrain and the unstable rock materials encountered. The highway was opened for traffic on schedule and within the budget established by the Utah Department of Transportation.

Although no lives were lost, the slide was still considered the worst natural disaster in Utah's history. It had buried a railroad, destroyed portions of two major highways, and flooded a village. Total direct cost of the damage was estimated at over $200 million.

After the lake had been completely drained the bottom surface was thick with residual mud and sediment. Because of this condition, as well as the understandable fear of another flood, few of the residents ever relocated in Thistle.

THE INVESTIGATION

The story of the Thistle landslide and the ensuing chain of events was widely publicized in both the newspapers and the technical society publications. At the time we wondered how long it would take for the forensic wheels to roll. Not too long. In early 1984 one of the authors received a telephone call from an attorney representing URC. The attorney asked if we would consider acting as expert witnesses in a matter involving the Thistle landslide. It turned out to be an intramural issue: URC was being sued by D&RGW in a dispute over how the repair bill should be split between the two railroads.

Our initial response to the invitation was that the dispute seemed to be a legal matter and did not require technical input. However, further discussion revealed that URC wanted an objective and scientific study of the entire event, particularly regarding causation. The scope of the study would include a complete geologic, geotechnical engineering, and hydrologic investigation to determine the history and causes of the slide, as well as an analysis of how the slide might have been prevented or mitigated. We accepted the assignment. Accordingly, a geotechnical engineering consulting firm with whom we frequently worked was employed to round out the forensic team. A summary of the main facets of our joint investigation follows.

Effect of Rainfall

The year 1983 was Utah's wettest in the twentieth century. The previous year also had greater than average precipitation, and the sequence represented the wettest 2-yr cycle in the state's history. These storm conditions wrought havoc throughout Utah, causing extensive damage and record-breaking losses from landslides, debris flow/mud flow, and flooding. Utah and the Salt Lake Valley/Wasatch Front had suffered from other record rainfall and snowfall periods, but property damage and financial losses in these previous episodes had been less, in part because of lower population and real property values. The 1982–1983 precipitation caused many ancient Utah landslides to move for the first time in centuries. For most of these slides the only apparent cause of reactivation was an increase in subsurface pore pressures due to excessive precipitation.

In researching rainfall records we found a USGS circular that showed cumulative departure from average annual precipitations at Salt Lake City from 1875 to 1982. We projected the departure curve through 1983. Figure 4–4 depicts the cumulative departure from the average of 15.63 in. (39.7 cm) for the 108-yr period.

It is noteworthy that, while the 1981–1982 and 1982–1983 rainfall seasons represented the steepest increase departure from the average, two other periods of high rainfall were extremely close. Figure 4–4 shows that the 1975–1976 season and the period from 1906 to 1910 were comparable in rainfall magnitude to the period that triggered the landslide. This fact led us to the early conclusion that the rainfall pattern, while unusual, was not sufficiently deviant to make it unforeseeable or to be able to characterize it as an act of God.

Even without subsurface explorations there was no question among the many geologists who examined the subject slope, before and after the failure, that the cause of the massive and sudden reactivation was saturation. Slopes that are saturated to ground surface are at least twice as unstable as equivalent slopes in which the water table lies beneath the potential failure surface, all other geologic conditions being equal.

Previous Geologic Observations

We discovered in our literature search of university archives and other sources that the Thistle landslide had been well documented historically in terms of geologic reports (see Fig. 4–5 for a geologic map of the area). For

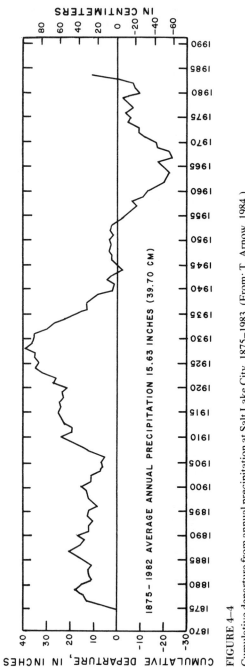

FIGURE 4–4

Cumulative departure from annual precipitation at Salt Lake City, 1875–1983. (From: T. Arnow, 1984.)

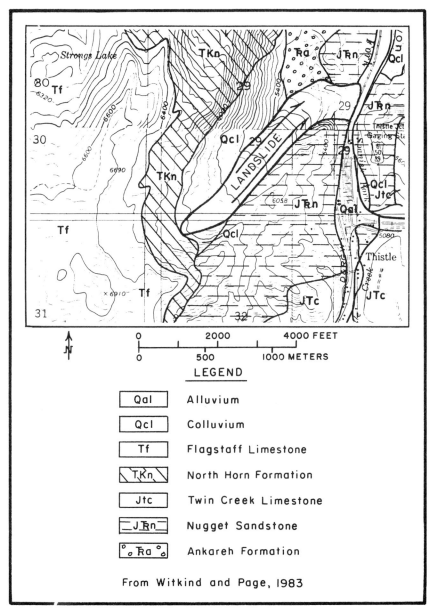

FIGURE 4–5
Geologic map of the Thistle area.

instance, a 1952 Brigham Young University Master of Science thesis by P.R. Peterson contained the following relevant comments:

> The main streams, though normally confined to their channels, become destructive torrents during periods of heavy runoff as witnessed by damage during record flows in the spring of 1952. . . .
>
> Slumping is common on the steeper hill slopes. This phenomenon is particularly evident at the head of the basin west of Thistle. . . .
>
> As previously noted in the report some difficulty in describing certain formations and locating some contacts was experienced due mainly to slope wash. The area is generally undergoing active erosion. This is evidenced by gullying and normal stream erosion and by an unusual amount of slumping and sliding on certain steep slopes. The latter phenomenon is very pronounced near the head of the basin west of Thistle.

In 1957 another Brigham Young University Master of Science thesis by R.R. Rawson noted that "slumping is common on steeper hillsides, particularly in Spanish Fork Canyon and the railroad cut directly north of Thistle. West of the railroad cut, Tertiary sediments fill an ancient drainage channel in the Nugget sandstone. . . . During periods of heavy rains, slumping presents recurring problems of railroad clearance."

A 1962 report from *BYU Geology Studies* by J.F. Rigby contained this observation:

> Active creep of North Horn Flagstaff sediments through a gap in the Navajo Sandstone hogback north of Thistle, in Sec. 28., T. 9 S., R. 4 E., forms a hummocky slope area on the west side of the canyon. The highway was moved from the west to the east side of the canyon, in spite of obvious difficulties, because of the constant movement of the creep mass. The old highway grade can be seen ending against the rolling topography at the southern margin of the slide area.

In 1968 and 1976 came two additional reports from the author of the 1962 studies which contained the following excerpts:

> The younger Tertiary rocks are soft and have flowed down into Spanish Fork Canyon on the south margin. This produces the hummocky rolling landscape visible above the railroad cuts in the sweeping bend of the canyon. A long tongue of landslide debris extends down through a gap in Navajo Sandstone on the west of the canyon 1.5 miles (2.4 km) ahead, north of Thistle. . . .
>
> To the right, across Spanish Fork Canyon, can be seen a hummocky surface which is the result of creep and landslide of soft sedimentary rocks which overlie the tilted rocks of the canyon bottom. . . . The formations in the bottom dip upstream 15 to 20 degrees but the overlying formation is essentially horizontal. Additional evidence of this pronounced angular relationship, termed an angular conformity, can be seen up canyon east of Thistle.

In 1971 the following Thistle landslide data and analysis were published in the *Utah Geological and Mineral Survey Bulletin:*

Previous work:	map, Metter, 1955; description, Rigby, 1962; map, Hintze, 1962
Type:	complex slump and debris-flow
Dimensions:	width 4,000 ft (1200 m) at head, 1000 ft (300 m) in middle, 900 ft (270 m) at toe; length 8000 ft (2400 m), thickness 50 ft (15 m), volume 25 million (20 million m³) cubic yards
Elevation:	crown 6800 ft (2040 m); head 6500 ft (1950 m); toe 5100 ft (1530 m)
Rate of movement:	very rapid to slow
Slope exposure:	northeast
Vegetation:	sagebrush and scrub oak
Geologic setting:	conglomerate, sandstone and red shale of the North Horn Formation of Cretaceous-Tertiary age, which is overlain by Tertiary limestone, shale and sandstone of the Flagstaff Formation and conglomerate and red beds of the Colton Formation, also of Tertiary age
Causes:	poorly consolidated, argillaceous nature of the North Horn Formation
Correlation:	numerous slides have occurred, dating from late Pleistocene until very recently
Geomorphic age:	early youth to maturity, as shown by successively younger slides headward

This slide well illustrates repetitive or retrogressive movement. Continued instability in the head region is maintained by the formation of the main scarp after each episode of movement. Subsequent triggering effects produce succcessive landslides each shorter and smaller than the preceding because of the reduction in slope and available unstable material.

Other references or communications that contained data of value to us were:

1. Utah State University Conference Proceedings (1984). The author B.N. Kaiser stated that movement of the Thistle landslide had been documented over a period of years. His records showed that the Thistle landslide had moved essentially every winter and spring when there was significant, above average rainfall and snow melt.

2. Notes on Thistle Landslide (1983). The author, a D&RGW division engineer, stated that the first record of movement or slide at the toe of the Thistle landslide was in the early 1900s. He also noted that the D&RGW had removed several thousand cubic yards of landslide debris from the tracks in

the 1950s. Additionally, he remembered that the track had "humped-up" and moved east. His estimate of such movement was 5 ft (1.5 m) up and 3 ft (0.9 m) laterally (eastward). The author also noted that the previous division engineer had informed him that similar types of movement had occurred in the 1940s and 1950s.

Relying on these references, other data, and field reconnaissance, it was apparent to us that the Thistle landslide was geomorphically youthful prior to the 1983 failure and fit the criteria necessary to be classified as an active landslide. Minor reactivation during the past century was adequate warning that large-scale reactivation was a possibility. Available data suggest that had a stability analysis been run prior to the April 13, 1983, failure, it would have been apparent that a failure was imminent.

Field Tests

After failure, a host of geologists and geotechnical engineers, representing various interests, studied the landslide to determine the mode and cause of failure. Test borings drilled into the toe and lower and middle portions of the landslide by a private engineering and testing company in 1984 were made available to us. These data included rather complete cores of borings, boring logs, piezometric records, tiltmeter data, and physical parameters determined by laboratory testing.

We reviewed these data and selected one particular boring as the best representation. From this boring, samples of what was determined to be slide plane or failure surface materials were extracted and transported to our geotechnical engineer's laboratory. Multiple samples were reviewed, and samples from depths of 155 and 206 ft (46 and 62 m) were tested. Both these samples were of shear zones, the lower one being from within that portion of the boring that the USGS had logged as the base of the ancient debris flow/landslide. We had picked slide planes at 155, 206, and 312 ft (47, 62, and 94 m); however, suitable specimens were not available for testing from the zone of 312 ft (94 m). Visual preview of the cores strongly suggested that the slide plane gouge logged at 312 ft (94 m) was as weak, if not weaker, than the materials at 206 ft (62 m).

The cores had been stored in a dry and sometimes hot building for 7 mo (December 1984–July 1985) before the specimens were obtained by the authors. Thus the materials were quite dry. Saturation of the sample for 7 days under test confining loads brought the specimens to a moisture content somewhat below the suspected moisture content at the time of failure. Therefore, although some disturbance of the sample occurred through drying and rewetting, we believed that the test data showed strength parameters that were higher than, but still reasonably representative of, strength parameters along the slide planes.

Depth		Angle of internal friction cohesion (degrees)	Cohesion	
(ft)	(m)		(psf)	(Pa)
155	46.5	17.5	480	23,040
206	61.8	22	210	10,080

Seepage had been reported for years as issuing from the toe of the slide adjacent to the railroad tracks. It was therefore of interest to find, if possible, the piezometric head present in the slide mass and if any variations existed. To identify hydrostatic pressures within the slide mass, multistaged piezometers had been installed in each of the borings as it was completed (Fig. 4–6). The piezometers showed that there was a significant rise in the hydrostatic head in and near failure planes. More specifically, the general hydrostatic profile was well below the ground surface (10–80 ft; 3–24 m) but in close proximity to the slide planes; the hydrostatic head was found to be as much as 65 ft (20 m) above the ground surface. This artesian phenomenon verified further that the deeper slide plane, below 310 ft (93 m), was a valid active slide plane.

Simple slope-stability anlayses were performed using the "method of slices" to determine the general range of the factor of safety, with varying hydrostatic heads acting on the assumed failure surfaces. These analyses were based on the simplified assumptions of a uniform hydrostatic head on the assumed failure surface relative to the ground surface. However, the results, which are presented in Figure 4–7, simply confirmed what we already knew—that failure was predictable with the preexisting conditions. In other words, a readily achievable lowering of the hydrostatic pressure on the slide planes would have produced a factor of safety greater than 1.0.

Mitigation Study

It was obvious that the only economically feasible way to have prevented or mitigated the landslide substantially would have involved a systematic program to lower the water table. To insure its effectiveness it would have been necessary for the dewatering system to have been operative prior to the 1982–1983 rainfall season or certainly at least before March 1983.

In our judgment dewatering would have consisted of three components:

1. A grading operation to fill and provide positive surface drainage for two sumps or small lakes near the top of the slide. Stagnant water in these sumps caused considerable infiltration to the groundwater regime. In addition, numerous other smaller sumps or ponds should have been eliminated by means of minor grading.
2. Channelization of two major natural watercourses on the slide surface in

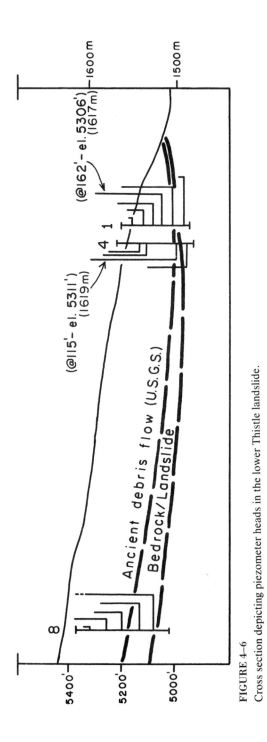

FIGURE 4–6

Cross section depicting piezometer heads in the lower Thistle landslide.

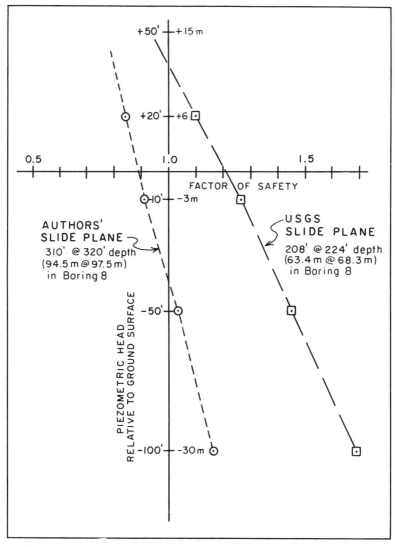

FIGURE 4–7
Stability analysis.

order to hasten rainfall and snow-melt runoff to Spanish Fork River and thereby decrease infiltration
3. Extraction of water from the subgrade by means of horizontal gravity drains or hydraugers

Figure 4–8 diagramatically illustrates the proposed surface drainage improvements and the hydraugers (cross sections shown in Fig. 4–9). The

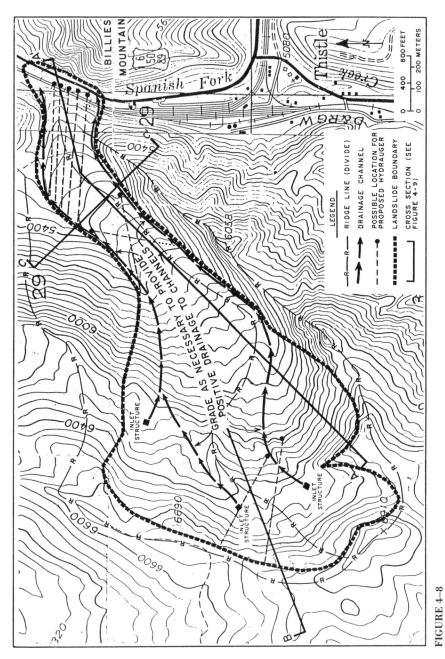

FIGURE 4–8

Proposed preslide mitigative drainage system.

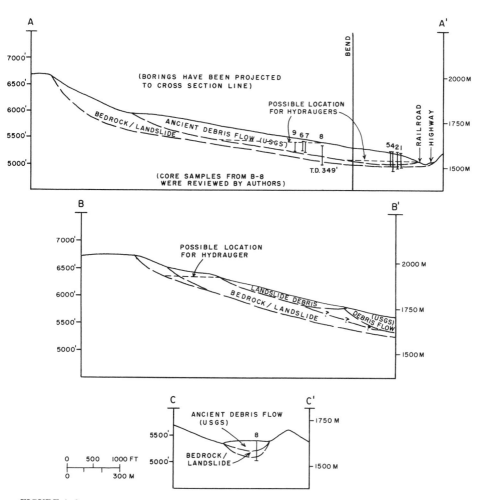

FIGURE 4–9
Cross sections shown in Figure 4–8.

proposed Gunite channels have a bottom 3 ft (0.9 m) wide, 2:1 side slopes, and a capacity of about 250 ft³/sec (7.5 m³/sec). Grouted rock drop structures also were included in the schematic design to decrease erosion due to the excessively high velocities that would be generated in the channels.

Nine hydraugers 2 in. (5 cm) in diameter were proposed, varying in length from 1800 to 2200 ft (540 to 660 m). From our experience such hydraugers placed in similar materials have developed flows as high as 50 gal/min (3.15 liter/sec) each.

Total construction cost estimates for the proposed improvements were:

Grading and channels	$430,000
Hydraugers	180,000
Total	$610,000

If the channels were not Gunite lined the total cost would have been reduced to about $375,000 but would have required considerably greater mainte-nance. The proposed scheme would have entailed minimal maintenance.

It was our opinion that the proposed surface and subsurface drainage scheme would have lowered the water table sufficiently to abate slide reac-tivation. Data obtained from the piezometers showed that the piezometric head was as much as 65 ft (20 m) above ground surface and averaged at least 25 ft (7.5 m) above ground elevation in January 1985. Since these data were obtained from borings drilled after failure, it was reasonable to assume that the piezometric head was higher at the early stages of failure, prior to ground-water exiting from the toe of the landslide as spring flow.

Landslide Dimensions

The landslide attained a length of over 6800 ft (2040 m), and its thickness (or depth to the basal slide surface) ranged from 200 to 250 ft (60 to 75 m) per USGS data, and possibly up to 350 ft (105 m) per the authors' studies. The width of the slide mass ranged from about 700 ft (210 m) in the lower extremi-ties to over 3600 ft (1080 m) in the headscarp area. The majority of the slide varied in width from 900 to 1100 ft (270 to 330 m).

Conclusions

The Thistle landslide had been clearly identified as a youthful (or active) landslide, its geometry had been determined and recorded by 1967, microac-tivity had been recorded from the turn of the century, and the general rela-tionship of precipitation to slide activity had been recognized by many geolo-gists. Data were obtainable by way of standard practice subsurface exploration, sampling, and testing (some were substantially obtained after failure) to determine

  (a) Location of the postslide activity hydrostatic pressures within the slide mass
  (b) Effect of precipitation on the water table and groundwater regime
  (c) Strength parameters for the slide plane materials
  (d) Effect of water table on the stability of the slide mass

Analysis of these data indicated that a lowering of the water table by utilizing surface and subsurface drains would have prevented failure and would have mitigated the inherent landslide hazard except for minor slumping or sloughing at the toe. Cost of mitigation with minimal maintenance was estimated at approximately $600,000. This figure represents about 0.3% of the estimated total cost of the landslide. It should be pointed out that our conclusion regarding the effectiveness and cost of the dewatering system was disputed by some of the other experts who studied the proposed plan of dewatering.

THE LITIGATION

One might expect that a disaster of this economic impact would result in multiple lawsuits, countersuits, and appeals that would swirl within the judicial maelstrom for years. Undoubtedly, if the event had occurred in California a monumental volume of legal activity would have ensued. In the more restrained and less litigious atmosphere of Utah, only two lawsuits were filed and only one ever came to trial. The first suit, as previously noted, was an internal dispute between the two railroads that was settled prior to trial. The second involved 15 of the Thistle residents versus D&RGW, a case that did reach the courtroom.

In 1984 D&RGW brought suit against URC for payment of its share of the maintenance, repair, and new construction costs. D&RGW had spent about $40 million for the repair and realignment work (of which $30 million was ultimately recovered by insurance). Under temrs of its lease arrangement, URC was to pay a fixed amount for normal maintenance but to share equally any costs caused by an act of God. From 1912 to 1983, fulfilling this obligation was no problem for URC. However, paying half of $40 million was another matter. An obligation of that amount might have bankrupted such a small company.

When URC refused payment, D&RGW sued. URC attorneys predicated their defense on two main themes:

1. That the realignment and tunnel work did not constitute maintenance and repair in the generally accepted use of the term as it applies to leases. Their argument, based primarily on legal precedents, was that the repair work represented a capital improvement expenditure, albeit not a voluntary one. In addition, URC would argue that it should have been consulted and given the opportunity to take an active role in the planning if it was expected to bear such a heavy financial burden. After all it was only the lessee, and the tracks and right-of-way belonged to D&RGW. If this distinction could be promulgated in court, URC would probably be responsible only for a minor

portion of the costs. All these arguments would have been of a legal nature and would not have involved our expertise.

2. That D&RGW exercised imprudent judgment in not attempting to mitigate the landslide prior to the winter of 1982–1983. The argument would surely have been put forth that D&RGW personnel knew of the existence of the slide and had performed considerable realignment and slide debris removal work since the 1950s. Therefore the landslide was not an act of God. D&RGW would probably have countered this allegation by arguing

(a) That there were numerous known landslides along this alignment and to have attempted to mitigate each would not have been economically feasible

(b) That over 99% of the subject landslide was not on railroad property. To obtain access rights for the construction of dewatering improvements would have been beyond the normal realm of what a railroad would be expected to do.

(c) That there was no assurance that any reasonably priced dewatering system would have been successful in preventing or substantially mitigating the landslide

After both sides had taken depositions from the opposing expert witnesses, it had to be obvious to the D&RGW lawyers that we were prepared to testify that

1. The subject landslide was more recognizable and had been more active than any of the other current slides adjacent to the route.

2. By a dewatering system estimated at about $600,000 the landslide could have been averted or, at the very least, mitigated to the point where the river could have been kept open and tunneling not required.

3. A cost-effective geologic investigative program could have been instigated to assess the potential hazards posed by all ancient adjacent slides and particularly this one.

4. The slide represented a foreseeable disaster; the only question was when it would happen.

After the D&RGW geology consultants had been deposed, it was evident that their testimony would be along the following lines:

1. The landslide was purely a natural disaster, an act of God, unpredictable in its magnitude without elaborate and expensive pre-1983 testing.

2. The landslide was the result of two of the wettest rainfall seasons in Utah annals.

3. Any reasonably priced dewatering system would not have been adequate to prevent or substantially mitigate the landslide.

If the case had come to trial the central issue would undoubtedly have been whether the landslide was a predictable event or an act of God.

Another dispute would have focused on the relative effectiveness of the dewatering system we had proposed. Before-and-after runoff infiltration calculations were fairly easily estimated based on rainfall records and applicable runoff coefficients. The quantitative effect of the hydraugers would not have been as easy to estimate as the surface hydrology. We would certainly have had to show a fairly substantial lowering of the aquifer due to the channels and hydraugers.

The case had a happy ending. A few weeks prior to time of trial, D&RGW and URC officials and their attorneys were able to arrive at a settlement satisfactory to both parties. We were not privy to details of the sealed compromise settlement, but both sides appeared reasonably happy. They had been friendly, cooperative neighbors for too long to become involved in what probably would have been a bitter and divisive trial. If D&RGW had won, URC might have been driven into bankruptcy proceedings, in which case D&RGW would probably have realized only a fraction of its claim. On the other hand, if URC had won, D&RGW would have had to bear the entire cost alone.

The other case, *Berrett et al. vs. D&RGW,* was filed in 1986 and did not reach the courtroom until 1989. Berrett was one of 15 plaintiffs who were former flooded-out residents of Thistle.

In early 1989 we were contacted by the attorney for D&RGW to see if we would act as his expert witnesses in the case. Our first reaction to the request was, is this a conflict of interest? Upon reflection we decided it was not, even though we had represented their opponents in the first case. Our reasoning was that

1. The first case had been settled.
2. We would present the same technical data, evidence, investigative results, conclusions, and opinions as we had planned for the first case.
3. The central issue was different.

It seemed that the attorney for D&RGW might have been gambling by using us, knowing we would be subject to hard, in-depth questioning regarding our deposition testimony in the previous suit. It was D&RGW's position that the central issue in the case was not whether mitigation was possible, but rather whether the railroad cut near the toe of the slope was the cause in fact of the disaster. Other than to review all the data and material from the first case, we did nothing in addition to prepare for the trial.

The plaintiffs' attorneys had alleged negligence on the part of D&RGW in creating a cut at the base of the slope for its tracks, which negligence was a proximate cause of the ultimate failure in 1983. The evidence, particularly the geomorphologic and aerial photographs, showed that this was not a regressive failure caused by the cut, but rather that the mountain, which had been moving for years, became oversaturated and was going to fail. This failure would have occurred irrespective of the railroad cut and would have occurred in the same magnitude.

The jury trial lasted about eight days. After deliberating for only two hours, the eight-member jury found in favor of the defendants. There simply was no credible evidence presented by the plaintiffs to convince the jury that the railroad cut was the cause in fact of the Thistle landslide. In fact the over-whelming evidence presented by the railroad clearly demonstrated that the cut did not add to the disaster.

It seems a shame that the residents received little or no financial aid for their losses other than about $5000 each from the federal disaster program. Although some residents carried homeowners' insurance, the carriers gener-ally refused payment on the grounds that their standard policies specifically excluded damage from flooding. But was this flooding in accordance with a strict technical interpretation? Or was this a back-up of water due to a constriction that might have been prevented? This interesting question was never explored in this disaster. In Chapter 7 we discuss a strict definition of flooding as it relates to homeowners' insurance policies and its effect on that particular litigation.

THE POST MORTEM

The Big Rock Mesa landslide, described in the previous chapter, resulted from a host of interrelated factors involving political considerations, greed, dishonesty, incompetence, and bureaucratic inertia. The Thistle landslide cannot be attributed to any of those human deficiencies.

D&RGW officials certainly appear to have acted in good faith from the time the landslide was first recognized. Their performance from April 13, 1983, and thereafter was exemplary. The sole criticism of D&RGW might be termed a lack of understanding of the importance of geologic expertise in evaluating the potential hazards of ancient landslides. In this respect most civil engineers and even some geotechnical specialists have neither the train-ing nor the experience to recognize a latent landslide disaster, to advise as to appropriate test procedures, or to recommend mitigation measures. The ideal scenario in Thistle would have been for D&RGW and the Utah Department of Transportation to have jointly employed geologic and geotechnical consul-tants to conduct an investigation complete with soil borings and laboratory testing. Certainly by the late 1970s, when the tracks had already been dis-placed a total of 5 ft (1.5 m) vertically and 3 ft (1.0 m) horizontally, D&RGW should have initiated such a study. Utah county and state officials (besides the department of transportation) also should have been involved at this point since protection of the community of Thistle was in their province.

Once an evaluation of the hazard had been completed, a joint decision could have been made as to the mitigation to be done and how the mitigation

LANDSLIDE RANKING

GREATEST RISK	ACTIVE	Motion is visible and/or measurable
	INCIPIENT	Minor change in internal water condition would accelerate movement
INCREASING RISK	INACTIVE	Change in internal or external factors may initiate movement
	STABILIZED	Stabilized by buttress fills, alluvial fill in canyon, etc.
LEAST RISK	ANCIENT / STABLE	No change in geomorphic shape, reduced hydrostatic levels within the slide, and soil or rock shear strength gain with time.

FIGURE 4–10
Landslide ranking.

costs should be shared. The magnitude of the movement was not a surprise to those geologists who had studied the landslide, even without benefit of borings or testing.

It was the opinion of some of the experts that there were just too many landslides adjacent to the railroad to consider mitigation and that to attempt to mitigate would not be cost-effective. We think the hundreds of significant landslides could easily be studied and catalogued using a classification methodology that would list them as (a) moving (active); (b) incipient; (c) quasi stable; (d) stable in current environs; and (e) stable unless major changes occur (see Fig. 4–10). The landslides could also be classified according to the extent of damage their reactivation would cause relative to topography, proximity of structures, etc.

It is estimated that slides in classificaitons (a) and (b), which are so located as to pose an immediate and serious hazard, would be less than 10% of the total. From our experience the cost of mitigation for the relatively few landslides that would fall into this category would represent a sound investment.

5   ABALONE COVE LANDSLIDE: MITIGATION INSTEAD OF LITIGATION

THE EVENT

Portuguese Bend Landslide

Any account of the 1974–1986 Abalone Cove landslide on Los Angeles County's Palos Verdes Peninsula must start with a brief history of the adjacent Portuguese Bend landslide, whose destructive movement began in 1956.

The disaster at Portuguese Bend is generally considered the granddaddy of Southern California slides and was significant in several respects:

- The devastating damage it caused
- The possible relationship to subsequent landslides in the area, including the subject slide at Abalone Cove
- The wide scope of the geologic and geotechnical engineering investigations by experts for both the plaintiffs and the defendants
- The greater than usual influence of the expert testimony on the outcome of the litigation
- The application of the legal doctrine of inverse condemnation on a theretofore unprecedented scale

Geologists, geotechnical engineers, and attorneys, some of whom were not yet born at the time of the Portuguese Bend landslide, debate today the merits of the forensic procedures used, the quality of the expert testimony, the strategy and tactics of the attorneys, and the verdict itself, which was to have profound effects on the Abalone Cove case.

Figure 5–1 depicts the general location of both landslides on the Palos Verdes Peninsula. Politically the peninsula contains several separate incorporated communities, with the remaining area under the jurisdiction of Los Angeles County. At the time of the 1956 Portuguese Bend slide, the land encompassed by the two slides was unincorporated. When the 1974 Abalone Cove slide commenced, the area was within the city limits of Rancho Palos

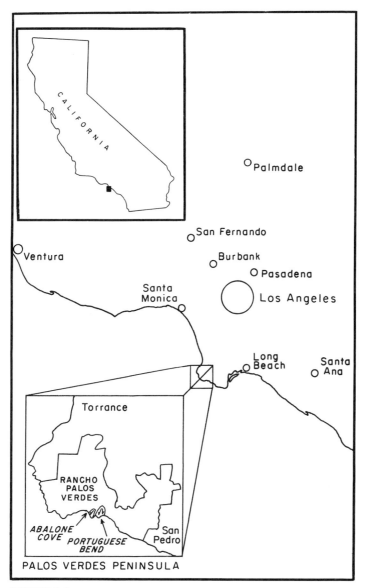

FIGURE 5–1

Vicinity map of the Portuguese Bend and Abalone Cove landslides.

Verdes. The slide would provide a major administrative and financial problem for the small city.

The topography of the peninsula is generally hilly, with the steepness of the terrain ranging from gently rolling to moderately steep. A bluff exists along the Palos Verdes coastline, varying in height from about 100 to 200 ft (30 to 60 m). Maximum elevation of the Palos Verdes hills is about 1500 ft (460 m).

Land use on the peninsula is predominantly single-family residential on large lots, many of which have panoramic ocean views. Palos Verdes is an affluent region where the residents zealously guard their privacy, their pristine environment, and their rural lifestyle. Much of the peninsula is zoned to permit horses and stables. Riding trails abound.

The Palos Verdes hills have long been recognized by most geologists as a region of unstable and marginally stable land. Unmistakable evidence of historic landsliding can be found everywhere, particularly along the south-facing slopes.

Figure 5–2 delineates the Portuguese Bend development as well as that at Abalone Cove. Significant topographic items shown on the map include the general surface contours, the approximate landslide boundaries, the watershed limits, the canyon watercourses, and the roads.

By 1956 there had been over 100 homes built in the Portuguese Bend area, about half of them southerly of Palos Verdes Drive South, a main road serving the peninsula, and half to the north. At the time no sanitary sewer system existed in the area bounding Palos Verdes Drive South. Plumbing in all the homes was connected to individual private septic systems, generally consisting of septic tanks and seepage pits.

Noticeable movement of the Portuguese Bend landslide southward toward the ocean began in late August 1956. Within 6 mo the movement had attained a maximum rate of about 1 in. (2.5 cm) per day. Over the next 4 yr the rate of movement subsided at an irregular pace until it reached about 3/8 in. (0.95 cm) per day in 1961. Since that time the slide movement has continued to diminish but has never stopped completely. The inches and fractions of incremental movement accumulated to a total of about 100 ft (30 m), although movement within the mass was not uniform.

By 1961 all 102 homes lying within the slide boundary, as well as the Portuguese Bend Riding Club, were totally destroyed. In addition, 52 homes north of Palos Verdes Drive South and west of Portuguese Canyon were either destroyed or seriously damaged. Since 1961 many additional houses have been impacted to some degree. Palos Verdes Drive South has been in a constant state of repair from 1956 to 1988, when an elaborate dewatering system began to greatly reduce the slippage rate. All underground utilities in or adjacent to the highway have been replaced above ground with steel pipe and flexible-type couplings. The utilities have also been a chronic maintenance problem.

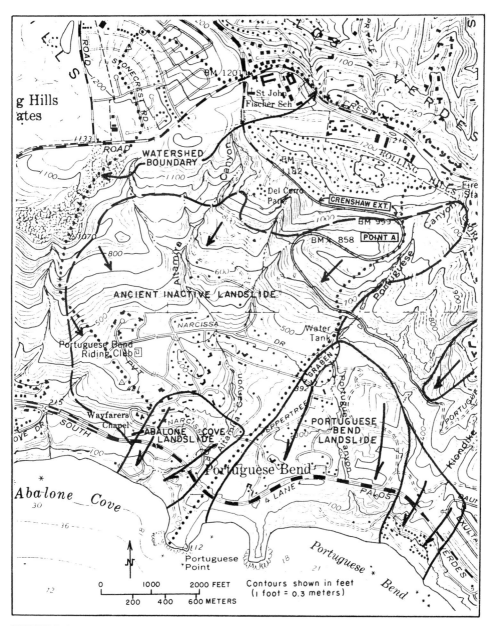

FIGURE 5–2

Site map of the Portuguese Bend and Abalone Cove landslides.

Low-angle oblique aerial photograph looking northerly across the Portuguese Bend and Abalone Cove landslides. Pronounced headland in center of photo is Inspiration Point; to its immediate left is the Abalone Cove landslide. Photo courtesy of Geo-Tech Imagery, Oceanside, California.

Figure 5–2 delineates the approximate present boundaries of the Portuguese Bend landslide as well as the subject Abalone Cove slide. There are now approximately 200 surface acres (80 ha) within the Portuguese Bend slide limits. Depth of the slide mass approached a maximum of about 190 ft (57 m), and its total weight was in excess of 35 million ton (35.3 million metric ton).

A key element in the litigation that followed the landslide was the grading of Crenshaw Boulevard as shown in Figure 5–2. The extension of Crenshaw Boulevard through the Palos Verdes Hills to Palos Verdes Drive South had been contemplated by the county since 1935. An alignment determined by field survey was established in 1937 but met hostile resistance from the landowners because it would dissect two existing residential areas. A new route was negotiated that bypassed the residential areas but traversed land owned by a major development corporation. The corporation donated the right-of-way but, to protect its own interest in terms of future development, played a significant role in determining the precise alignment. Although the alignment was agreed on and made official by 1940, World War II interrupted and it was 1950 before grading began.

Most of the work was performed by detention-camp personnel under the direction of the road department's maintenance division, which was not an uncommon procedure at the time. As the grading progressed, frequent changes were made in alignment and grade. These revisions were generally due to a combination of three circumstances: (*a*) unforeseen geologic conditions such as adverse bedding planes that necessitated flatter cut slopes, thereby creating additional earth to be used as fill; (*b*) the economic desirability of balancing cut and fill; and (*c*) the accommodation of the development objectives of the property owners.

There was no consideration given in the design or construction of the road to the fact that it would be crossing an ancient landslide. Existence of the ancient landslide had been known for years by local geologists. Detailed information was contained in a 1946 report by the USGS. Its existence was apparently recognized by county employees as well as some of the landowners and developers and their consultants. However, it is understandable why the ancient slide was ignored by all concerned:

- The year was 1956, a time when geologic hazards as nebulous as ancient landslides were just not made part of the road planning process.
- Even in retrospect it is difficult to conceive what could have been done other than to drastically reroute the road to circumvent the slide. Considering the abundance of other ancient landslides on the peninsula, it would have been difficult, if not impossible, to weave an alignment among the ancient slides that would fulfill the road's traffic function, satisfy grade and alignment design criteria, and meet the demands of the property owners.

Road grading was in progress but was immediately halted when the landslide began. By that time the earthwork had been essentially completed to point A as shown in Figure 5–2. The total volumes of road cut and fill within the slide boundaries (including that of the slope easements and adjacent disposal areas) was approximately as follows:

Cut   =   37,000 y^3 (29,600 m^3)
Fill   = 175,000 y^3 (140,000 m^3)
Net fill = 138,000 y^3 (110,400 m^3)

Thus there was superimposed near the top of the slide mass an additional weight of about 186,000 ton (189,000 metric ton). This additional surcharge, representing about one-half of 1% of the total weight of the landslide mass, was to play an important role in the litigation.

A total of 129 lawsuits were filed in the wake of the Portuguese Bend landslide. Eventually, in 1961, all of them were consolidated into a single mammoth nonjury trial. Charges and cross-complaints of liability could be summarized as follows:

1. The corporate landowners charged the county with liability based on negligence, creation of a dangerous and defective condition, and inverse

condemnation by reason of construction work in Palos Verdes Drive South and Crenshaw Boulevard.

2. The county cross-complained against the developers and homeowners contending that development was the cause of the slide, particularly the introduction of sewage effluent through the septic systems and infiltration from landscape watering.

3. The residents cross-complained against the developers charging fraud and negligence in the sale or lease of property in a known hazardous area and for participating with the county in the construction of Crenshaw Boulevard.

4. The local water company sought recovery of damages to its water systems under an all-risk insurance policy. The carrier denied liability claiming the water company concealed facts as to the risks involved.

Construction of Crenshaw Boulevard proved the focal point of the case against the county. Negligence could not be established, so the plaintiffs relied on inverse condemnation. Simply stated, the plaintiffs' lawyers argued that if the county had not undertaken the grading of Crenshaw within the slide limits, the slide would not have occurred.

Experts for the plaintiffs, including several of the nation's most prestigious geotechnical engineering and geology consultants imported from afar for the purpose, testified that the additional surcharge near the head of the slide produced sufficient extra driving force to set the slide in motion. Emphasis was placed on the supposition that the slide started at the head. If true, the location of the road fill was more significant than if the fill had been placed closer to the toe of the slide.

Experts employed by the county testified that the added weight of the road fill was negligible in comparison to the entire mass. One county geotechnical consultant even produced stability calculations showing that the added weight was beneficial in that it increased the frictional resistance force slightly more than the driving force it created. In reading the trial transcripts it seemed to us that the plaintiffs' attorneys and experts did a more credible and well-executed job of presenting their data than did the county. County experts did not appear to adequately quantify or relate the effect of the septic systems and landscape irrigation on the rise in the water table. This was evident from the fact that the judge, in evaluating the effect of the septic systems, made two flawed observations:

1. He compared total infiltration from septic systems with total rainfall rather than with rainfall infiltration.

2. He compared total weight of the effluent infiltration with the weight of the road-fill surcharge, not taking into account other detrimental effects of the effluent such as loss of strength of the clays in the slide plane and effects of hydrostatic pressure.

Several months after the trial ended the judge handed down a lengthy and thoroughly deliberated decision. The bottom line was that he found no negligence but did find the county liable in inverse condemnation for the grading of Crenshaw Boulevard and that the road-fill surcharge was the proximate cause of the landslide. Based, in our opinion, on lack of evidence, the county's position that the seepage of septic system effluent and irrigation to the groundwater regime was the proximate cause (and probably the actual cause) of the slide was given little or no credence.

Little did anyone suspect that the judge's decision would assume such weighty importance more than 20 yr later.

Abalone Cove Landslide

This landslide, like the one at Portuguese Bend, represented the reactivation of an ancient slide. Unlike Portuguese Bend, the reactivation ws gradual, progressing in small increments dating back to 1974.

Initial movement appears to have started at the coastline and involved a mass having a surface area of about 9 acres (3.6 ha) seaward of Palos Verdes Drive South. Few details are available regarding progress of the sliding between 1974 and 1978. Movement continued either very slowly or intermittently, with cracks extending as far inland as Palos Verdes Drive South by 1976. Maximum horizontal displacement during the 4-yr period was about 4 ft (1.2 m).

Movement of the area inland of Palos Verdes Drive South commenced in late April or early May 1978, following one of the wettest rainfall seasons on record: 29.61 in. (75.2 cm) fell in the area during the 1977–1978 season as compared to an annual average of 11.38 in. (28.9 cm).

The rainfall season of 1979–1980 was also abnormally high. During the one week of February 13 to February 21, 1980, 7.75 in. (19.2 cm) fell. This high-intensity storm resulted in a greater than usual proportion of infiltration and caused another surge of movement.

A dewatering program consisting of newly drilled (dewatering) wells had been initiated in March 1979. The wells were successful in drastically slowing movement, even arresting the northeasterly portion of the slide. However, a temporary resurgence of movement occurred in the autumn of 1980, when two dewatering wells were shut down for repair of damage caused by earlier slide movement. In addition, 1980 had also been a year of abnormally high rainfall. Another high-intensity 9-day storm of 4.29 in. (10.9 cm) in March 1981 instigated further movement. By the end of 1981 the slide seemed nearly stabilized, at least temporarily, with its rate of movement slowed to less than 0.01 in. (0.025 cm) per day.

Unfortunately the 1982–1983 season brought a record 30 in. (76.2 cm) of rainfall, resulting in further movement. Three comparatively dry seasons followed, and, coupled with the continuing and improved dewatering, the slide exhibited apparent zero motion in late 1986. By that time the landslide had damaged about 25 homes and put at least 90 others at risk.

Residents filed lawsuits against several public agencies targeting, in particular, Los Angeles County. The two principal allegations against the county were:

1. The Abalone Cove landslide represented an expansion of the earlier Portuguese Bend slide. Therefore, since the county had been judged liable in inverse condemnation for the Portuguese Bend slide by dint of its work on Crenshaw Boulevard, such liability would simply carry over to Abalone Cove.
2. The Portuguese Bend slide created a sizable graben or tear-away depression at the head of the slide. It was the residents' contention that for nearly 30 yr the graben had been intercepting and impounding surface runoff and that most of this impounded water seeped underground in the direction of the Abalone Cove slide. This subsurface diversion, it was charged, was a proximate cause of the Abalone Cove slide.

Graben at head of Portuguese Bend landslide. Whether storm-runoff accumulation in graben affected Abalone Cove landslide was a major item of controversy.

THE INVESTIGATION

Several months after the lawsuits were filed against the county the authors were employed by a member of the county counsel's staff to conduct an investigation into the causes of the Abalone Cove landslide. A hydrogeologist also was hired to supply expertise in groundwater migration. Eventually the county decided to contract the litigation to a private law firm, which inherited us as expert witnesses.

There were no strict limits applied to the scope of our investigation. However, it was immediately apparent that we would be focusing our attention on the following questions:

1. What evidence existed that the two landslides were part of the same larger slide?
2. Would the Abalone Cove slide have occurred in the absence of the Portuguese Bend slide? If so, would it have occurred within approximately the same time frame?
3. What quantities of water infiltrated the groundwater regime in the Abalone Cove slide mass during the period between the two slides?
   (a) From rainfall?
   (b) From septic systems?
   (c) From irrigation?
   (d) From the Portuguese Bend landslide graben, if any?

Our investigations were stopped when the decision was made to mitigate rather than to litigate. The tentative results of our studies at that time can be summarized as follows:

- Although the two slides appeared to be components of the same ancient landslide, we found that the Abalone Cove slide moved independently of the Portuguese Bend slide from 1974 to 1986. The most compelling evidence is that they were physically separated by a narrow strip of land. Surveys of monuments on the strip showed only very slight movement from 1956 to 1987. The fact there was any movement at all was probably indicative that the two slides were riding piggyback on an extremely deep, very slow moving slide, whose horizontal limits may approximate or even be greater than the ancient slide.
- Employing similar techniques as with the Big Rock Mesa slide discussed in Chapter 3, we estimated infiltration to the Abalone Cove groundwater regime during the period from 1956 to 1983:

From rainfall	=	9,765 acre ft	(1201 ha m)
From septic systems	=	845 acre ft	(104 ha m)
From irrigation	=	137 acre ft	(17 ha m)
Total		10,737 acre ft	(1322 ha m)

- Our analysis of the effect of the Portuguese Bend graben during this same period:

Runoff to graben   = 752 acre ft   (93 ha m)
Evapotranspiration = 113 acre ft   (14 ha m)
  Net volume
    infiltration   = 639 acre ft   (79 ha m)

If the investigation had continued it was the intent of our team hydrogeologist to drill several observation wells: some between the graben and the Abalone Cove slide and others south of the graben in rough conformity with the surface contours. He then planned to wait until water had ponded in the graben after a storm to introduce a red dye, or isotope tracer. If water pumped from the wells periodically over the next year disclosed traces of the dye or isotope, it might have been possible to make some qualitative judgments about whether some of the water was migrating toward the Abalone Cove slide. This procedure had apparent weaknesses. First, there would have been no rational way of quantifying the migration in each direction. Second, since the rate of migration would have been difficult to estimate, there would have been no assurance that sufficient time had elapsed for the dyes to appear. Finally, there may not have been a sufficient number of wells to accurately track the migration routes through the complex, highly fractured rock structure. How much time and money would have been warranted to pursue such elusive conclusive evidence?

To put the matter into perspective, if one makes the extremely unlikely assumption that all the infiltration from the graben ended up in the Abalone Cove slide groundwater regime, the percentage of such migration as against the total of all infiltration would have been

$$\frac{639}{10,737 + 639}$$

or about 5.6%. This would have compared to about 7.4% from the septic systems.

THE MITIGATION

Our pretrial investigation had proceeded intermittently for several years when all the parties involved in the pending litigation came to a momentous agreement. They decided that money to be spent on a lengthy and expensive trial would be better spent on mitigation of the problem. Thus in 1984 a panel of six mutually agreed on technical experts, including one of the authors, was appointed by the various involved attorneys to analyze the landslide and to consider feasibility, methodology, and a cost estimate of stabilizing the ground in the Abalone Cove area, with the overall goal of developing a means of preventing future ground movement and protecting the homes from further damage.

This same panel, which was empowered by the court to oversee the mitigation, has convened on a regular basis to this time. Three of the members are engineering geologists, two are geotechnical engineers, and one is a hydrogeologist. In addition to their consulting credentials, each was or had been teaching at the college level.

The original panel was given a budget of $192,500 to conduct investigations, initiate corrective action, and present recommendations for stabilization. Because the panelists agreed that they had insufficient geologic and engineering data to assess the feasibility of stabilization, half of the budget was allocated for geologic field investigation and engineering computer analyses. The remaining budget was spent on interpretation of the results of the field studies, meetings, and consultations and on the preparation of a report summarizing the panel's findings and recommendations. By the end of December 1984 the original panel had determined that stabilization was both possible and feasible. A technical report was drafted by one of the members and edited, approved, and signed by all the others.

Field investigations undertaken by the panel included the drilling of five new borings in the slide area. Piezometers and slope indicators were installed, and soil samples and rock cores were obtained. Although drilling conditions were difficult, the geologic and engineering data necessary for stability analyses and for the subsequent engineering and cost estimate for various stabilization plans were obtained. The installed instrumentation will continue to provide useful information in the future.

Detailed logs of the borings were analyzed by the geologists. New information was obtained on the relationship of the Portuguese Tuff and the depth of the slide plane. A new contour map and cross sections showing the depth and location of the slide plane were prepared and used as a basis for stability and engineering evaluations.

A comprehensive analysis of past information on groundwater within the slide supplemented by new data from piezometers was conducted by the members of the panel with expertise in groundwater hydrology. Particular "compartments" into which the overall slide seems to be divided were identified, and the efficiency of the existing dewatering system evaluated in this light. Locations for new wells with potentially high efficiency were identified. Depths of past, present, and probable future water-level fluctuations were estimated for purposes of stability analyses.

Design of landslide stabilization measures and prediction of their future performance required the development of a stability model, a mathematical representation of slide behavior, which adequately represented the past behavior of the slide and which therefore could be counted on to predict the future performance of a stabilization program.

Two different modeling techniques were used by the geotechnical engineers for cross-checking purposes. The modeling techniques included the latest developments in the field of geotechnical engineering, and when combined with the improved data obtained in the geologic and hydrologic investigations, provided a more accurate and reliable assessment of landslide behavior than had been available in the past.

In general the stability analyses showed that distinct improvements in the factor of safety could be obtained by several different methods, but the improvements resulting from each method were of a magnitude such that a combination of several methods would be necessary to attain a satisfactory factor of safety.

Engineering feasibility assessments and cost estimates were prepared for various landslide stabilization techniques, including an improved program for control of surface and subsurface water in the Abalone Cove beach area. Cost estimates were developed largely on the basis of experience with similar projects and were intended to be sufficiently accurate for budgetary planning purposes only. Preparation of detailed plans and specifications for the stabilization and a relatively accurate engineering cost estimate based on such plans and specifications were not a part of the panel's scope of work.

The panel did not discover any geologic or hydrologic conditions that differed in major ways from previous generalized interpretations. However, the newly obtained data provided the minimal factual basis necessary to evaluate the effectivenss and cost of stabilization measures.

It became clear as stability studies progressed and first attempts to prepare cost estimates for stabilization were initiated that local conditions, notably depth to water and slide plane, were quite variable and significant from the standpoint of stabilizing the moving ground. The borings in the slide toe area showed that the elevation of the slideplane varied from 34 ft (10 m) below sea level at the west side of the slide to 84 ft (25 m) below sea level on the east side. This was a surprising discovery, as the depth to the slide plane in the toe area was found to be much deeper than previously thought. Borings upslope of the active slide showed that the depth to the base of the slide was closer to 150 ft (45 m) than to the approximate 100 ft (30 m) previously estimated.

The groundwater hydrology studies showed the importance of viewing the slide mass as a complex series of subsurface groundwater compartments with variable rates of percolation into the slide from upgradient and within the slide itself. The location of and interconnections between major compartments formed the basis for the location of eight additional dewatering wells. It was expected that these wells would capture and exhaust water that escaped the initial system, thereby lowering pressures (and water tables) that were destabilizing the slide. This element of the stabilization program was given first priority.

The panel did not attempt to evaluate or reach agreement on the physical causes or exact mechanism of the landslide movements. However, the panel members agreed that the existence of weak geologic materials, historical and prehistorical sliding, introduction of water into the slide mass from both natural and human-associated sources, and coastal erosion were all significant factors in the ground movements that began in the early or mid-1970s and continued to 1986.

Modeling studies confirmed a landslide mechanism involving a very large mass of earth—an area of about 92 acres (37 ha) and 150 ft (45 m) thick weighing on the order of 33 million ton (33.3 million metric ton). The mass was found to be sliding at a rate varying from inches per year at the top of the slide to feet per year at the toe of the slide on a weak clay layer that is gently inclined at an angle of about 6 degrees. During wet periods the driving forces activating the slide were several hundred thousand tons more than the resisting forces. Stabilization of the slide required that the resulting "resistance deficit" be eliminated in the future. In theory this could be accomplished several different ways. However, the latest geologic data and modeling studies indicate that the particular circumstances at Abalone Cove limited the most practical solutions to two: increasing the resisting forces by controlling water in the slide, and adding resisting forces by building a stability berm/buttress at the beach area. See figure 5-3 for descriptions of these stabilization methods.

The panel believed that groundwater movement in the slide complex could be controlled by an effective combination of the elimination of domestic sewage by constructing a sewer system, development of surface and storm drain facilities that prevent infiltration of storm waters originating both within and outside the slide area, and the operation of an improved and expanded permanent groundwater pumping system. The complete water control program would involve mandatory domestic sewers and roof drains for all houses, land grading to fill and seal depressions and cracks in the ground, construction of debris basins, construction of a major storm drain facility in Alta Mira Canyon, and expansion and improvement of the existing groundwater pumping system.

The geologic exploration program carried out by the panel provided information that, if adequately confirmed, points to a toe berm as an additional stabilization method that would supplement or possibly be a substitute for some elements of the water control system. The apparent presence of the following geologic conditions provided a basis for designing a counterweight berm that would retard or prevent further movement of the toe: (a) failure at the toe of the slide, which rises relatively steeply from an average depth of 60 ft (18 m) below sea level to where it crops out along the beach; (b) strong rock dipping seaward in the surf zone and beyond the slide toe; and (c) moderately strong (low cohesion but moderately high friction) displaced bedrock above the slide plane at the toe.

Stability analyses indicated that, in theory, the weight and dimension of the berm could be configured to counteract the combined forces of both the relatively active sliding below Palos Verdes Drive as well as the additional forces generated by the slower moving and more stable area above Palos Verdes Drive.

Preliminary analyses indicated that a counterweight berm weighing about 700,000 lb/ft (95,000 kg/m) of beachfront could have an effectiveness equivalent to a reduction of 20 ft (6 m) in slide water level created by a water control program. A berm of this magnitude would consist of fill 50 ft (15 m) high extending seaward roughly 100 ft (30 m); for example, an artificial 100-ft (30-m) seaward extension of the seacliff and beach.

Adequate protection against erosion by rock or other such protection in conjunction with gentle sloping of the fill would be necessary. Possible sources of fill could conceivably include waste fill from construction or demolition as well as dredging or barged-in waste material. Excavated earth from higher parts of the slide (e.g., from the "knobby" topography between the beach and Palos Verdes Drive) for placement in the beach area, would probably be the most economic source of fill.

Following the preliminary studies more detailed stabilization studies showed that the net effect of such a mass redistribution from higher to lower ground could stabilize the entire area. With a total earthwork volume of about 300,000–500,000 yd³ (240,000–400,000 m³), basic construction costs would be a minimum of $2–3 million . To this must be added the costs of beach-erosion protection to safeguard the counterweight berm/buttress from the wave action typical of the Southern California coastline, and a cofferdam for protection and dewatering during construction. Total costs could reach $10 million.

Clearly a coastal earthwork program would encounter regulatory and permitting problems (California Coastal Commissions, U.S. Army Corps of Engineers, California Lands Commission, etc.), but at the same time would also offer the potential benefit of creating an attractive park environment with improved beach facilities. Of course design of anything more than a primitive stability berm would add to the cost. As of this writing the viability and cost of the berm are still being studied by the panel.

Recommendations

The panel was confident that the sliding could be permanently arrested if all of the stabilization methods (see Fig. 5-3) were put into place. The cost of the complete mitigation program would be at least $10 million. Significant savings might be achievable if a program of progressive treatment is undertaken and favorable results are attained early in the program. Most of these costs are for construction materials and services and do not include inflation, interest, or administrative or other overhead expenses. Neither do they include any fund allowance for future operation and maintenance of the improvements. The accuracy of these costs varies from ± 20% for water control measures to very rough approximations for less well defined items such as the toe berm.

The panel's studies indicated that the factor of safety achieved by implementing less than the full program is insufficient to provide the level of assurance expected in California residential land development or even equal to the level of assurance that would have been thought to exist if the slide had not become active in the first place.

However, the panel did not recommend that all elements of the program be simultaneously undertaken, even if funds were available to do so. As is the case in the treatment of a difficult medical problem, the appropriate engineering response is to undertake a course of progressive treatment of the condition, with the most radical and expensive steps held in reserve in case an early improvement is not forthcoming.

The panel members agreed that the simplest, most cost-effective, and most quickly attainable improvement in land stability would be realized by minimal grading and sealing of ground fissures and depressions and installation of more dewatering wells (items 2a and 5a in Fig. 5-3). Installation and operation of an extensive water-level and ground-movement monitoring system (item 2b) would provide an improved ongoing means of measuring the effectiveness of the dewatering and other stabilization measures. Funding of less than $1 million would be sufficient to establish and operate these dewatering and monitoring systems.

Although analyses indicate that it is likely that the recommended upgraded dewatering system would stop slide movement within a few months, the panel opined that the continued discharge of domestic sewage into the slide and the intermittent heavy stressing of this weak slope due to occasional storm-water infiltration, if left uncontrolled, will continue to pose an unacceptable level of hazard to the existing developments. Permanent control of these destabilizing conditions will require capital investment of up to $4.5 million.

Neither the cost nor the construction planning feasibility of the toe berm (item 6) can be stated with great certainty now, but the panel believes that the principle of the toe berm must be considered as a supplementary reserve method if all other measures should prove insufficient.

Again, the proposed monitoring system (item 2b) would provide a sounder basis for more definitive evaluation and design of toe stabilization measures if such measures are determined to be necessary or desirable.

In summary, the proposed progressive program of treatment could, under the most unfavorable circumstances, ultimately cost in excess of $10 million. An early improvement in the apparent stability of the slide, on the other hand, could justify a reduced program costing on the order of $5 million. However, the panel did not believe that a high level of confidence should be attached to a prospect of achieving these savings.

It should be noted that the prinicipal aim of the mitigation plan was to

STABILIZATION METHOD	OPERATIONS INVOLVED	FEASIBILITY	COSTS	
			INITIAL COST, $	ANNUAL OTM, $
1. Maintain existing dewatering system	Inspection, maintenance and replacement of pumps, wells; power supply	Established	0	?
2. Expand existing dewatering system	a. 8 additional wells	Feasible	140,000	20,000 ?
	b. Monitoring wells, slope indicators, evaluation		300,000?	25,000 ?
3. Domestic sewer system	Connect 90 domestic systems to County Sanitation District's main on Palos Verdes Dr. South	Feasible	1,340,000	? user charges?
4. Individual lot storm drainage	Gutters, downspouts, lot drains	Feasible	430,000	?
5. Storm Drainage (Alta Mira Canyon drainage area-approx 800 acres [324 ha])	a. Fill/seal existing depressions and fissures	Feasible-but needs further investigation	100,000	0
	b. Street improvements	Feasible	270,000	?
	c. Other storm drainage per Slosson memo. Lesser system also possible	?	1,910,000-2,410,000 (for full system)	?
6. Toe berm	Placement of about 50 ft. (15m) of fill, 300,000 cu. yds. (230,000 m^3) along beach, with associated shore protection, landscaping, etc. Most fill obtained between beach and Palos Verdes Dr.	Needs verification. "Shear key" toe stabilization is possible less costly method	5,000,000 reduce to slope protection?	?

FIGURE 5–3

Proposed stabilization methods (from Slosson and Havens, 1987).

ANTICIPATED EFFECTIVENESS		OTHER COSTS/BENEFITS	MANAGEMENT CONSIDERATIONS
IMPROVEMENT IN FACTOR OF SAFETY	CERTAINTY OF ACHIEVEMENT		
100 acre-ft.(12.3 ha-m) pumpage per year has reduced water level about 25 ft. (7.6m) average; equivalent to an improvement in FS of 8%	Established		Continued responsible management essential
Improve effectiveness of existing system by picking up water missed by current system	Certain	Rapid implementation and improvement of stability is assured	Continued responsible management essential
Up to 30 acre-ft (3.7 ha-m) per year reduction in infiltration for full system	Certain	Political and sanitation considerations ?	Presumably managed by County Sanitation District
Approx. 12 acre-ft (1.5 ha-m) per year reduction in infiltration for 90 homes	Certain		Requires some management
Under existing situation each 1" (2.54cm) of rainfall infiltration could cause roughly 2 ft. (0.6m) rise in water level. Produces 1% reduction of the safety factor.			Private property easements required. Management required. May require EIR
Approx. 4% improvement in FS. Fill volume and FS could be increased proportionally, e.g. 60,000 cu. yds (46,000m³) will produce 8% increase in FS	Possible significant regulatory and permitting issues	General improvement in beach and park access and facilities, with possible revenue production	Significant ownership and management issues would have to be addressed. Will require EIR and Coastal Commission approval.

regain the value of the property affected by the landslide as well as its usability. The proposed progressive mitigation measures should achieve this goal, making their cost justifiable.

The cost of mitigation, even if currently underestimated, will still be considerably less than the overall cost of litigation.

THE POST MORTEM

In some respects the efforts to mitigate the Abalone Cove landslide have been disappointing. The decision to mitigate was finalized in 1984, yet 7 yr later most of the improvements recommended by the panel are not in place.

The reasons for such tediously slow progress include the following:

- Since the slide had apparently been arrested using only dewatering wells there was a general feeling of complacency, at least not a sense of urgency, on the part of residents and government officials.
- The technical complexity of the problem resulted in an abnormally long period for the field work, laboratory tests, and analyses.
- Committees are rarely as effecient as an authoritarian administration.
- Some of the residents rebelled against certain mitigative items such as sewers, flood control facilities, and roof drains.
- The planning and design times for environmentally sensitive improvements such as the Alta Mira storm drain are lengthy.
- Cumbersome bureaucratic procedures on the part of the county prevent the releasing of funds on a timely basis.

Fortunately the past 6 yr have been relatively dry. In fact the past 5 yr have been termed drought years. A wet cycle might well have set the slide in motion again.

As of early 1991 the status of the improvements was as follows:

1. Expansion of dewatering system: Several new wells have been drilled and others will be added if deemed necessary.
2. Domestic sewer system: The final plans and cost estimates have been completed and construction is scheduled for early 1991 if funding has been made available.
3. Individual lot and street drainage: A limited number of lots and street areas have been graded to improve drainage. Palos Verdes Drive South through the slide area has been reconstructed to provide excellent surface drainage.
4. Alta Mira Canyon: The design documents have been completed and construction awaits funding.
5. Toe berm: This item has caused some controversy within the panel, and design parameters are still pending.

Prospects for Mitigation in General

Despite the mixed results of the Abalone Cove mitigation program we are confident that mitigation succeeds in many cases.

We have participated jointly or individually in well over 100 environmentally related forensic assignments. The disposition of the lawsuits stemming from these events is estimated as follows:

Litigated in court	60%
Settled prior to trial	38%
Mitigated rather than litigated	2%
(one besides Abalone Cove)	

Further, we estimate that of those cases litigated, at least 25% could have been effectively and fairly mitigated with a tremendous savings in both time and expense. Most of the cost savings would have accrued to the public in its dual capacity as consumer and taxpayer.

Why isn't mitigation used to a greater extent to remedy the wrongs caused by environmental disasters and failures? Because for mitigation to hold sway over litigation, at least some of the following conditions must be present:

- Attorneys must be willing to forego lucrative litigation fees for more modest fees involved in mitigation negotiations and administration. We have witnessed cases where legal fees alone (not including other court expenses, expert witness fees, etc.) have actually exceeded the estimated cost of mitigation.
- Attorneys for both plaintiffs and defendants must have some doubt about the outcome of a trial. If either side is supremely confident of winning in court, the opportunity to work out a mitigation solution is slim.
- A judge in sympathy with a mitigation approach and who can be persuasive with both sides during pretrial settlement conferences can be a strong factor in influencing a mitigative effort.
- Well-informed plaintiffs must exert pressure on their attorneys to pursue mitigation when circumstances warrant. Any rational approach by plaintiffs should weigh the advantages of mitigation (i.e., more rapid repair of the incurred damage, more timely action to prevent further damage, and quicker settlement of other monetary issues) against the present worth of a questionable successful court verdict that could be 5 yr down the road, not including possible appeal time. In other words, plaintiffs must fully understand the benefits of mitigation as opposed to the risks and rewards of litigation.
- If there are insurance claims involved the carriers, of course, must cooperate in the mitigation process.
- Above all there must be defendants or potential defendants who are willing to advance the funds needed to perform the remedial work, such as Los Angeles County in this case.

Obviously the odds are stacked against mitigation because of these requirements and because plaintiffs' attorneys are often loathe to surrender potentially huge contingency fees in favor of hourly fees earned in establishing and
administering a mitigation program. However, because of its innate fairness
and efficiency, every effort should be made to make mitigation work.

The Elusive Safety Factor

One of the conditions for approval to build in a known or suspected landslide
area is that the geotechnical engineer must submit calculations showing that
the postdevelopment ratio of resisting forces to sliding forces is 1.5 or greater.
Why 1.5? Because it appears to represent a reasonable compromise between
unfairly penalizing the landowner or developer and providing for errors in
stability calculations that, by their very nature, cannot be precise.

The degree of accuracy depends on how closely the assumptions on which
they are based reflect actual field conditions. Factors such as strength of the
earth materials, weight of the soil and rock, elevation of the water table,
frictional resistance of the materials, cohesion, passive pressure, and pore
water pressure will vary throughout the site or slide mass. Calculations need
to be based on valid samples of the material involved (or that may become
involved) and the *worst* assumed conditions. In the zeal to show a site safe
and stable this often does not happen.

An unfortunate joke recently making the rounds among geology and geotechnical offices goes something like this: Irate hillside developer to his
geotechnical engineer: "Why is it taking so long for you to secure a grading
permit from the county? Don't you realize how much money I'm losing each
day this project stands idle?" Engineer's reply: "We're working as fast as we
possibly can. We're trying to find a core sample on the site that will test out to
yield a 1.5 landslide safety factor. Be patient—we'll eventually succeed."

It would be nice to report that such conversations do not take place.
However, the high incidence of landslides in areas where favorable geotechnical and geologic reports have been submitted is testimony that such
dialogue is occurring all too frequently. Believe it or not a geotechnical firm,
in support of proposed development, is reported to have calculated a 1.5
safety factor for the Portuguese Bend landslide, even though the slide is still
moving, albeit slowly. By its very definition a safety factor of 1.0 represents
equilibrium. If the slide is moving it necessarily means that the safety factor is
less than 1.0.

In addition to the games played with safety factors there is another practice
by geologists that has led to many landslide failures—that is, ignoring or
proving wrong USGS maps that show the presence of landslides on properties being investigated for development. It is not that the USGS is infallible in

identifying slides, but it is a competent and objective organization whose findings should be heeded carefully.

Perils of Cityhood

When Rancho Palos Verdes divorced itself from Los Angeles County in 1957 and incorporated as a separate city, most of its citizens rejoiced. There were the usual valid reasons for celebration:

1. The potential for lower taxes made possible by a smaller more efficient government administration
2. More responsive services
3. Most important, the opportunity to control its destiny through its own land-use master plan, zoning regulations, and building codes

One of the disadvantages of deannexation that may have been overlooked at that time, and as seen may still be overlooked by Malibu residents, is the tremendous liability exposure posed by environmental-type lawsuits. Landslides in particular are inevitable in areas like Rancho Palos Verdes and Malibu unless there is a 100-yr drought.

Despite its unusually high residential tax base, it was rumored that Rancho Palos Verdes found itself in dire financial staits several years ago as the result of losing several landslide cases in which it found itself the "deep pockets"defendant. Unfortunately most smaller cities do not have the financial resources to carry more culpable defendants who have neither the assets nor sufficient insurance to pay their fair share of the award. To make matters worse, liability insurance for small cities has become practically unobtainable or, at best, prohibitively expensive.

Fortunately for Rancho Palos Verdes it was incorporated in 1957 and therefore played no part in the decisions that determined the alignment and grade of Crenshaw Boulevard. Otherwise it would probably have been stuck with a portion of the Portuguese Bend landslide verdict as well as part of the Abalone Cove mitigation costs.

Rancho Palos Verdes was recently able to secure a grant from the state for $3 million to spend on further mitigation improvements. Few communities could have exerted that much political influence.

Smaller cities located in environmentally hazardous regions must be particularly careful and conservative on the side of safety, not only in approving and inspecting private development but in the planning, design, and construction of public improvements. A particular cause of nightmares for public officials is the inverse condemnation specter. It is simply not enough to be careful, diligent, and conservative in order to minimize the risk of negligence or misconduct. As is cogently illustrated by the Portuguese Bend–Abalone

Cove legal parlay, a seemingly innocent public works endeavor can be extremely costly to public bodies and the taxpayers who support them. If the costs related to these two landslides would have fallen to Rancho Palos Verdes instead of being spread over the vast county taxpayer base, the city could possibly have been driven into bankruptcy. One wonders what would happen to a less affluent city with lower property values and less political clout at the state level.

Double Jeopardy

When the county lost the Portuguese Bend case, it was ordered to pay the landslide victims the fair preslide market value of their properties, including both house and lot. After most fire, flood, earthquake, and high-wind disasters there is still residual value in the land. After a Portuguese Bend–type landslide the land is not only generally worthless but can be a distinct liability.

It may be for this reason that the county chose not to buy or condemn most of the properties, despite paying full value for damages. This policy proved to be a mistake since some of the property owners are suing the county again, alleging this time that the county had not acted properly in stabilizing the slide. Exposure to such double jeopardy is simply the result of bureaucratic and political shortsightedness.

There were two compelling reasons for the county to take title to the destroyed properties: First, so it would be able to exert control over the land. Perhaps with ownership the county would have graded the land to fill the graben. Our studies indicated that this would not have been an exceedingly costly undertaking (about $100,000) and would have nullified what could have been a difficult allegation to refute—that at least some ponded runoff water from the graben migrated underground to the Abalone Cove slide mass. Filling the graben would also have aided long-term stabilization. Second, when a reasonable degree of stabilization had been attained, the land could have been converted to a nonstructural public use such as a park or a golf course.

Rancho Palos Verdes Building Moratorium

In September 1978 Rancho Palos Verdes adopted an ordinance that established a moratorium on the issuance of grading and building permits and on the processing or approval of environmental impact reports, conditional use permits, and tentative subdivisions or parcel maps within a region identified as "The Landslide Moratorium Area." This area encompasses both the active and the inactive portions of the Abalone Cove landslide.

Recently geologic and geotechnical engineering reports have been submitted to the city by a developer who wants to build estate homes and a golf course on what is essentially the unimproved portion of the Abalone Cove slide complex. The reports and plans are being processed for possible approval even though the mitigation measures are incomplete and the factor of safety has not yet been determined. One can only hope that the potential property values involved will not take precedence over sound engineering and geologic practices.

6 FLOODING WITHIN A FLOODPLAIN WHILE THE SUN WAS SHINING

THE EVENT

When one thinks of international problems along our Mexican border, what immediately come to mind are illegal immigration and drug trafficking. It is not often that the root cause of a major problem is simply the topographic and geologic features of the land—aided and abetted by a large dose of political and bureaucratic indecisiveness, procrastination, and just plain bungling. Such is the saga of the Tia Juana River Valley in the City of San Diego, which was racked with flood damage in 1980. Incredibly, much of the flood damage occurred in bright sunlight without so much as a falling raindrop.

To comprehend the reasons for and ramifications of the flooding, it is essential to know something of the history of floodplain management in the Tia Juana River Valley.

The Tia Juana River originates at the confluence of the Rio de Las Palmas and Arroyo Montanoco Rivers in Baja California, Mexico, about 11 mi (18 km) south of the border. From there it flows northerly through the City of Tijuana (the difference in spelling between river and city remains a mystery), picking up Cotton Creek, a principal tributary. From Tijuana it spreads into a broad floodplain in the United States, where it flows westerly to the Pacific Ocean.

The 1700-mi^2 (4420-km^2) watershed feeding the Tia Juana River, as well as other salient features, are shown in Figure 6–1.

Prior to 1930 there were no major dams or channel improvements within the watershed. Storms of large magnitude produced heavy flooding throughout the lower valley, including Tijuana. At that time the population of Tijuana was about 5000, compared with over a million today. In 1930 the Rodriguez Dam (and Reservoir) was built on the Rio de Las Palmas at a site immediately south of the confluence. Named for the governor of Baja California at the time, General Abelardo L. Rodriguez, it was designed to impound water for agricultural irrigation and to serve as the city's water supply, which had been dependent on wells. A secondary purpose of the dam was to provide some measure of flood protection. However, as Tijuana's population bur-

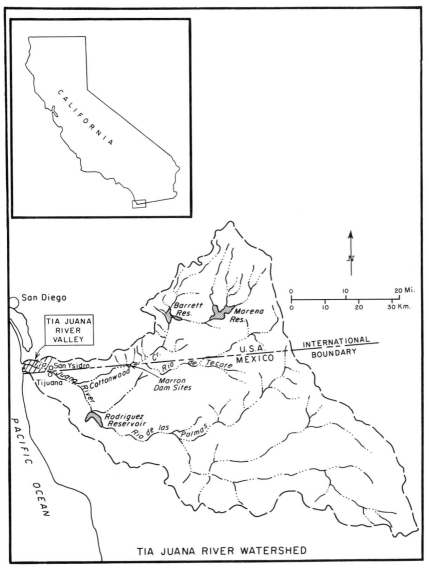

FIGURE 6–1
Tia Juana River watershed.

geoned, its increased water needs dictated a policy of keeping the reservoir as full as possible, releasing water through the floodgates only when it was felt that the structural integrity of the dam might be threatened.

Two other dams and reservoirs, Barrett and Morena, were later built on Cottonwood Creek within the United States. Their function was to supply water to the City of San Diego and, like Rodriguez, they eventually offered minimum storage for flood control.

In 1939 San Diego County and the City of Tijuana entered into an agreement creating the International Trunk Sewer Line, which carried raw sewage from Tijuana and the United States border community of San Ysidro to the ocean on the United States side of the border. No treatment of the sewage was provided. In 1944 the International Boundary and Water Commission was formally established by means of a treaty to deal with common problems regarding the Tia Juana River. The treaty preempted local and state laws because of the treaty-making powers of the federal constitution.

Despite the mitigating effect of the dams during light to moderate storms, Tijuana was subject to periodic floods from major storms, causing much damage and human suffering. At the United States end of the floodplain the basic land use was agricultural. While causing crop damage at times of severe flooding, the long-term net effect of such floods was beneficial since they deposited layer upon layer of rich silt to create fertile farmland.

To solve Tijuana's flood problem, Mexican authorities proposed a concrete-lined flood control channel 2.7 mi (4.3 km) long, terminating at the boundary. For many property owners on the United States side the Mexican channelization, if unilateral, would have been devastating. Due to loss of floodwater storage and decreased groundwater infiltration on the Mexican side as well as tremendously increased velocities at the border channel outlet there would have been severe erosion at the easterly end of the United States portion of the valley and excessive deposition at the westerly end.

A simple solution, at least technically, proposed by United States authorities, would merely have extended the channel about 6 mi (10 km) from the border to the ocean. The channel would, of course, have eliminated both the natural soil enrichment process and the groundwater recharge. In addition, the flood protection would have opened the door to other land uses. To compensate for the loss of groundwater, it was proposed there be a channel diversion to a 120-acre (48-ha) replenishment or recharge area. For many landowners this possibility was a mixed blessing. Their property values would appreciate, but their farms would be threatened. For environmentalists seeking to preserve the area as open floodplain or parkland, the proposed channel was an anathema. They feared the channel would have an adverse effect on the estuary as well as the natural processes in the valley.

After signing a formal agreement to extend the channel, the City of San Diego had second thoughts. All design work by the U.S. Army Corps of Engineers was stopped while the city studied four alternative plans. Under pressure from the International Boundary and Water Commission to make a decision so that the Mexicans could proceed, the city finally adopted the alternative proposed by the environmentalists: an energy dissipator at the border, a low flow channel to Dairy Mart Road, a 262-acre (105-ha) sedimentation area for a buffer, and a series of protective levees.

In accordance with their part of the agreement, the Mexicans started work on their channel in 1972, completing it in 1974. Mexican officials were not particularly concerned about the United States plan, since it posed no flood danger to Tijuana as long as the common border was protected by levees.

United States authorities recommended that the valley be given floodplain zoning, thus precluding urban development. To decrease velocities and spread the flow within the United States, the Corps of Engineers designed and built an elaborate state-of-the-art energy dissipator structure (Fig. 6–2).

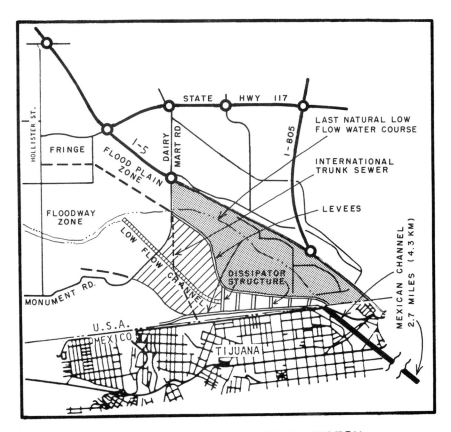

DISSIPATOR FLOOD CONTROL SYSTEM
TIA JUANA RIVER VALLEY

 SEDIMENTATION AREA

AREA PROTECTED FROM FLOOD HAZARD

FIGURE 6–2
Dissipator flood control system.

In addition to the dissipator, the low flow channel was proposed to be extended about 4000 ft (1200 m) to connect to the then existing natural low flow channel. To provide a buffer zone, a sedimentation area of 282 acres (113 ha) was purchased or condemned east of Dairy Mart Road.

To confine floodwaters to the undeveloped portion of the floodplain, earth levees were constructed along the north, east, and south sides of the valley. Mexican officials requested one opening be left in the south levee along their north boundary at "N" Street. The reason for such a request remains obscure, but the existence of this opening would prove to be a significant factor in the magnitude of the flood damage.

Meanwhile the international sanitary sewer outlet was replaced in 1966 with an interceptor sewer that traversed the valley, as shown in Figure 6–2, and that terminated in a City of San Diego outlet trunk line leading to a treatment facility.

The first real test of the dissipator came on January 30, 1980, a day of heavy rainfall over the entire watershed. On January 29 runoff into the Rodriguez Reservoir began to cause a rapid rise in its water-surface level. By 3:00 AM on January 30, reservoir capacity had been exceeded and Mexican officials, fearful of dam failure, opened up the floodgates. This decision apparently did not come easily. A federal official was flown in from Mexico City to supervise the operation of the gates. The reservoir's discharge, added to the runoff from the rest of the watershed, caused great flooding in the valley. Because there had been little preceding rainfall and because the reservoir discharge was relatively sediment free, the floodwaters were erosive in nature. Rainfall essentially stopped by the next day, but the Mexicans did not close the floodgates until almost 7:00 AM on January 31. Thus much of the damage occurred on January 31st when rainfall had ceased and the sun was shining.

The valley had only a 2-wk respite before a storm of even greater magnitude occurred. Rainfall began on February 15. Its intensity built up slowly so that by February 21 the peak flow at the international boundary was greater than the peak flow of January 30. This time the Mexican officials did not wait for the water to exceed maximum gage level. Instead they began discharging from the reservoir anytime the water reached a level 12 or 13 ft (3.6 or 3.9 m) below the maximum level for the reservoir. In contrast to the erosional effects of the first flood, the second flood predominantly caused high water and sedimentation damage.

The combined effect of the two floods created havoc in the valley.

- Erosion in the area of the interceptor sewer caused it to rupture. At about the same time the pipe ruptured, a Mexican sewage pumping station, located near the aforementioned opening in the south levee, broke down. Until it was repaired several weeks later, 15 million gal (56.8 million liter) of raw sewage per day flowed from Mexico through the opening and into the San Diego portion of the valley. Between the rupture and the overflow, most of the crops were contaminated and all were condemned by the State of California Department of Public Health.

- Many valley properties suffered erosion, others excessive deposition.
- Much farm equipment was either buried or washed out to sea.
- Some farm houses and other structures sustained water and mud damage.

When all the damage had been accounted for the total dollar amount was estimated at over $20 million. Fortunately there was no loss of life or serious injuries directly associated with the flooding.

THE INVESTIGATION

The authors became involved in a forensic investigation of the disaster in early 1983 as the result of an inquiry from the San Diego City Attorney's Office. We learned that the city was organizing a team of experts to defend itself against a lawsuit brought by valley property owners. Two of the other experts already on the team were consulting civil engineers who specialized in hydrology and hydraulics, while another was a geotechnical engineer. We accepted the assignment with the usual premise that our investigation must be both comprehensive and objective.

Our first step was to learn as much as we could from city officials and other public agency employees who had firsthand familiarity with individual aspects of the disaster. Early contacts included the City of San Diego Department of Public Works, the U.S. Corps of Engineers, and the International Boundary and Water Commission. Several field trips were arranged for us to examine the flooded area, the energy dissipator, the Mexican channel, and Rodriguez Dam. From the city attorney's staff we obtained copies of all the then current and relevant discovery information, and from a local television station we were fortunate to obtain a very useful videotape.

Most major environmental disasters and failures are well covered by the media. Newspaper and television reporters, along with their camera people, are often the first outsiders to arrive at the scene. Generally they are both skilled and objective. Thus any environmental forensic investigation should take full advantage of any media stories, photographs, and films.

Newspaper and television stations are usually very cooperative in providing copies of photographs and video film, much of which might not have actually been used in broadcasting the event. Sometimes the unused photographs or film clips may not be spectacular or of great human interest but may contain valuable investigative clues. Reporters, too, may have made observations or conducted interviews that never appeared in print or on television.

In researching media data it is important to be certain that all potential sources have been contacted. For instance, many of the smaller independent television stations are as good a source of information as network affiliates. The same can be said for the smaller local newspapers. However, a word of caution: Although the reporter, editor, and publisher ordinarily have no reason to be biased, their news sources may be.

In addition to the other sources of information, we were fortunate that one of the author's staff members was an eyewitness to the flood of February 21.

The principal allegations of the plaintiffs could be summarized as citing inverse condemnation in connection with

1. Construction and operation of the sewer main, which ruptured
2. Construction of the energy dissipator, which they allege did not function properly and which was oriented in the wrong direction
3. Construction of the Mexican channel, which allegedly increased velocity and volume of flood flow, prevented flood storage in Mexico and inhibited infiltration into the ground. (Even though the channel was built by the Mexican government, the argument by plaintiffs was that the existence of the treaty agreement made all parties responsible and that all improvements could therefore be considered to be a single project.)

The State of California also was named as a defendant since it had participated in the planning and financing of the project. However, because the federal government is protected by immunity from this type of action, the Corps of Engineers was spared.

Watershed Characteristics

Using USGS and Mexican topographic maps obtained through the International Boundary and Water Commission, we were able to establish a composite picture of the Tia Juana River watershed. The approximate area statistics are as follows:

Total watershed area $= 1700 \text{ mi}^2$ (4420 km^2)
Watershed area in Mexico $= 1245 \text{ mi}^2$ (3240 km^2)
Watershed area in the United States $= 455 \text{ mi}^2$ (1180 km^2)

Elevations within the watershed range from sea level to about 6000 ft (1800 m) in the mountains. The lower alluvial floodplain portion of the watershed contains about 100 mi^2 (260 km^2); approximately 60% is in Mexico and about 40% is in the United States. This fact would become important in determining the effects of the Mexican channel on decreased flood storage and ground infiltration and hence increased flooding on the United States side.

Flooding History

Mean annual rainfall in the lower Tia Juana River Valley is about 10 in. (25 cm) as contrasted to about 24 in. (61 cm) in the mountainous areas of the watershed. Approximately 70% of the rainfall occurs from December through March. Tropical storms of high local intensity, locally known as Mexican storms, occur during the summer months. However, because they seldom cover a large area and because so much of the rainfall is consumed in evapotranspiration and infiltration, they have never caused severe flooding in the valley.

Table IV, supplied by the Corps of Engineers, shows the dates and corresponding discharges of major floods in the Tia Juana River Valley from 1884 to 1963.

TABLE IV
Floods in Order of Decreasing Magnitude

		Discharge	
No.	Date	(ft^3/sec)	(m^3/sec)
1[a]	January 27, 1916	75,000	2250
2	February 1884	50,000	1500
3	January 1895	38,000	1140
4	February 16, 1927	25,000	750
5	December 1889	20,000	600
6	February 1891	20,000	600
7	February 7, 1937	17,700	530
8	March 24, 1906	16,000	480
9	December 26, 1921	15,000	450
10	February 22, 1941	13,800	410
11	February 23, 1944	11,100	330
12	April 11, 1941	10,400	310
13	March 15, 1941	8620	260
14	March 3, 1938	6760	200
15	February 21, 1914	5000	150
16	March 17, 1942	2770	83
17	December 24, 1940	2700	81
18	January 13, 1949	2600	78
19	January 7, 1944	2500	75
20	December 23, 1944	2100	63
21	March 13, 1938	1600	48
22	March 12, 1918	1600	48
23	February 18, 1932	1500	45

[a] The flood of 1916 is said to have destroyed most of the improvements in the valley. Two lives were lost, and the excessive deposition ruined the land for years.

The Corps of Engineers also computed an estimate of flood return frequencies for the valley:

Discharge		Approximate
(ft³/sec)	(m³/sec)	return frequency (yr)
2000	60	3
5000	150	5
20,000	600	12
30,000	900	20
50,000	1500	50
75,000	2250	100
135,000[a]	4050	335

[a] Standard project flood.

None of these estimates take into account the dampening effect of reservoir storage. However, it seemed reasonable to assume that, since the dams were used primarily for water supply, there was little storage available at the height of the major storms. The return frequency of the 1980 flood was apparently about 20 yr. This fact was of interest in putting the magnitude of the flood in general perspective and in relating it to the design of the energy dissipator.

Geologic Considerations

A thorough geologic investigation was undertaken with particular emphasis on the geologic history of the Tia Juana River Valley. It appeared extremely important to ascertain whether the low flow thread of the river had always followed the same course through the floodplain and, if not, to what extent its course had changed throughout the years.

Following is a generalized geologic overview of the lower Tia Juana River drainage:

1. The shape of the land (geomorphology) in and around the Tia Juana River Valley has been determined largely by alternate rise and fall of sea level with respect to the land (the alignment of the valley has been influenced by faults as discussed below).
    (a) As the result of uplift and subsidence of the land
    (b) As the result of raising and lowering of sea level

2. Areas above sea level at any given time in the past were subject to erosion by streams that over a short period of time, produced relatively narrow ravines or canyons.

3. Areas at sea level at any given time in the past were subject to planation (coastal erosion), during which wave-cut terraces formed.

4. Areas below sea level were protected from the erosive action of waves and running water and were thus sites of sediment accumulation deposition.

5. Much if not all of the present alluvium in the Tia Juana River Valley is younger than the last low stand of sea level, which occurred about 18,000–20,000 yr ago. Sea level at that time fell to approximately 420 ft (126 m) below present sea level on a worldwide basis.

6. The Tia Juana River Valley, as currently observed, is only the upper part of a much deeper valley now filled with sediment. Well data from the State of California Water Resources Division logs reveal that this ancestral valley was about 100 ft (30 m) deeper than the present valley and was essentially flat bottomed and floored by bedrock (San Diego Formation).

7. The valley was probably cut into bedrock during the last low stand of sea level. The depth of alluvial fill is significantly less than the sea-level minimum of 420 ft (126 m) below present, which suggests that the planation of the valley bottom took place as the result of stream erosion. Lateral broadening of stream valleys is a common process, involving shifting of a stream channel from side to side within its valley as well as deposition related to stream braiding.

8. An alternative explanation for the presence of a relatively flat bottom in the ancient Tia Juana River Valley is that flattening occurred as the result of wave action during a previous high stand of sea level. This process would be similar to the process that produced other elevated flat surfaces in the San Diego area. The most likely time of valley flattening is the Sangomon interglacial period, which saw as many as five high stands at sea level ranging from 30 to 125 ft (9 to 37 m) below present during a period of about 60,000 yr from about 60,000 to 120,000 yr ago.

9. One of the most prominent land forms in the Tia Juana River Valley area is the slightly elevated flat surface that lies along its north side. This surface, called the Nestor Terrace, extends from the ocean on the west to Otay Mesa on the east. It has an average height of about 25 ft (7.5 m) above the floodplain of the Tia Juana River.

10. The Nestor Terrace is about 120,000 yr old and consists of sediment deposited on a wave-cut platform.

11. Worldwide sea level approximately 120,000 yr ago was about 20 ft (6 m) above present. This close agreement with the elevations of the Nestor Terrace indicates that there has probably not been much uplift or subsidence since that time in this area. However, it should be noted that the Nestor Terrace has been broadly warped as well as offset in the San Diego area in general, so the coincidence of elevations may be fortuitous.

12. Other terraces south of the Tia Juana River are older than the Nestor Terrace.

    (a) One terrace has been dated at 200,000 yr and lies at slightly more than 29 ft (9 m) above present sea level. Sea level 200,000 yr ago was more than 160 ft (48 m) below present.

(b) A second terrace, called the Lindavista Terrace, lies as high as 300 ft (90 m) above present sea level in the same general area and has an age of about 1 million years (early to middle Pleistocene). Elsewhere this terrace has been observed at highly variable elevations. At the mouth of the Tia Juana River it occurs nearly 100 ft (30 m) below sea level. At Otay Mesa it occurs at about 500 ft (150 m) above sea level.

13. The variable elevations of the older terraces indicate that significant uplift and subsidence have occurred since about 1 million years ago. Most uplift and subsidence occurred prior to 120,000 yr ago, probably owing to fault offsets.

14. Uplift and subsidence have been accomplished by both tilting and faulting in the Tia Juana River area.

15. Most of the Tia Juana River floodplain is bounded by mapped faults that have had an important influence on the orientation of the valley.

(a) The southern extension of the La Nacion fault zone forms a northwest–southwest-trending boundary along which Otay Mesa has been uplifted north and east of the river.

(b) Two known faults south of the Tia Juana River form a boundary, parallel to the one on the north, along which Spooners Mesa has been uplifted.

(c) The faults just mentioned are stepped downward toward the river on both sides of the valley, forming what is essentially a tectonic trough or graben that has controlled the alignment of the river valley and floodplain.

(d) The northwest trend of the valley in this area parallel to the previously mentioned faults is in marked contrast to streams to the north, which are oriented transverse to the northward extension of the La Nacion fault.

16. The Tia Juana River, like other rivers to the north, is presenting an alluvial cycle whereby it is filling its valley with sediment rather than actively eroding.

(a) The thickness of the alluvium is estimated at 50 to 100 ft (15 to 30 m) in the shore areas and up to 200 ft (60 m) offshore.

(b) The deposit consists of soil underlain by river sediment that includes predominantly sand and silt in the upper part and coarse gravel with cobbles and boulders in the lower part.

(c) The alluvium consists of channel, bar, backswamp, and generally fine grained overbank deposits interlayered with each other as the result of normal channel migration within the valley.

(d) The general decrease in coarseness toward the top of the deposit probably represents rise of base level with the last major rise in sea level. This would have caused a decrease in the ability of the river to carry coarse material. It also may have been affected by a change in adjacent topography to more subdued terrain. There was, of course, a dramatic decrease in velocity as the river entered the essentially flat Tia Juana River Valley.

With this general overview as background, we obtained about 20 topographic maps of the valley dating back to 1849. Sources for the older maps included the National Archives in Washington, D.C., the California State Archives in Sacramento, and local universities. Using each of these maps, we plotted the historical course of the flow thread of the river as it traversed the valley. These courses are shown in Figure 6–3.

It was obvious that there had been extreme changes of directions of the meandering flow lines through the years and that, left entirely to nature, there would be constant future changes. Not surprisingly these changes occurred at times of peak river flow, when high velocities cut new channels quite easily in the loose alluvial sediment.

At this point, even without hydrologic or hydraulic input, we were convinced that the orientation of the energy dissipator was not contrary to natural flow.

It was also apparent that, in 1980, the first storm's maximum flow eroded the floodplain into channels. The peak flows in the second storm were then more channelized and attained higher velocities.

Hydrology

Hydrologic studies were conducted by all members of the forensic team. One of the firms ran computer programs (HEC-1 and HEC-2) for flood storage and water surface based on various scenarios. We depended on these computer hydrology calculations for much of our analysis.

We were particularly concerned with the computer-run water-surface elevations downstream of the dissipator structure as they actually existed during the 1980 flood versus what those water-surface elevations would have been if there had been no flood control project at all (i.e., no Mexican channel, no United States dissipator structure, no low flow channel, and no levees). The results indicated that the water surface would have been insignificantly lower, in the magnitude of 0.1–0.3 ft (3–9 cm), without the project. Velocities also would have been slightly lower.

The reasons for these surprising results were apparently twofold:

1. The hydrograph for the storm was radically altered because of the huge discharge from Rodriguez Reservoir. Thus, with no project the effect of flood storage within Tijuana, which would have increased the time of concentration and decreased the peak discharge (Q), was nullified by a broadened hydrograph peak.

Location of low flow stream bed from historic maps dated:

1869 □□□□□□ 1941 ×××××××
1882 ∘∘∘∘∘∘∘ 1943 —·—·—
1883 ▵▵▵▵▵▵▵ 1953 —··—··
1890 ———— 1967 ●●●●●●
1904 — —

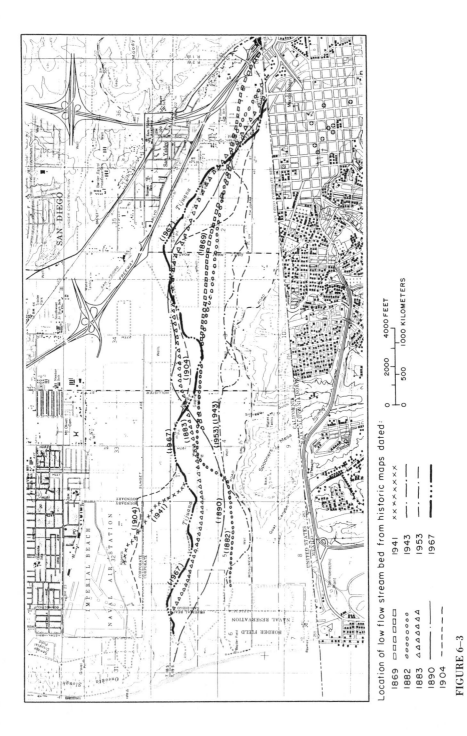

FIGURE 6–3

Historical low flow courses.

2. The dissipator structure was at least partly successful in reducing velocities and spreading the flood flow back into the natural pattern.

Figure 6–4 shows the hydrograph for the actual conditions, and Figure 6–5 shows the daily averages and peak discharges at the international boundary.

Energy Dissipator

When it was decided to abandon the idea of extending the Mexican channel to the ocean, the Corps of Engineers was given the task of planning and designing an alternative flood control project. The general concept was to build a transition from the end of the Mexican channel into an energy dissipator structure and then to control the lateral flow at the terminus of the structure by means of levees. The project was patterned somewhat after a similar structure in Walnut Creek, California.

During the planning phase, officials of the Corps of Engineers were anxious to dispel the notion that the project would provide flood protection to the properties downstream from Dairy Mart Road. In fact, the district director of the Corps wrote in a letter to a local congressman: "The purpose of the structure is to control flows emanating from upstream improvements in Mexico by reducing their velocity and directing them into the natural flood-

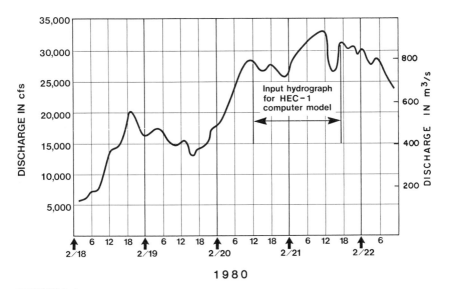

FIGURE 6–4
Storm hydrograph (from Leedshill-Herkenhoff, Inc.).

TIA JUANA RIVER DISCHARGES
AT THE INTERNATIONAL BOUNDARY

	DAILY AVERAGE DISCHARGE		INSTANTANEOUS PEAK			DAILY AVERAGE DISCHARGE		INSTANTANEOUS PEAK	
	cfs	m³/s	cfs	m³/s		cfs	m³/s	cfs	m³/s
JAN 1	3	0.1			FEB 1	4,300	122		
2	3	0.1			2	444	13		
3	3	0.1			3	2,060	58		
4	3	0.1			4	4,650	132		
5	3	0.1			5	408	12		
6	3	0.1			6	295	8		
7	4	0.1			7	280	8		
8	4	0.1			8	273	8		
9	54	1.5			9	266	8		
10	53	1.5			10	245	7		
11	156	4.4			11	232	7		
12	135	3.8			12	225	6		
13	55	1.6			13	258	7		
14	28	0.8			14	738	21		
15	14	0.4			15	3,450	98		
16	9	0.3			16	4,440	126		
17	11	0.3			17	5,550	157		
18	70	2.0			18	11,900	337	20,700	586
19	68	1.9			19	15,600	442	17,200	487
20	51	1.4			20	24,400	691	28,900	818
21	28	0.8			21	29,100	824	34,000	963
22	18	0.5			22	22,200	629	30,300	858
23	15	0.4			23	3,650	103		
24	14	0.4			24	2,250	64		
25	13	0.4			25	2,225	63		
26	9	0.3			26	2,600	74		
27	9	0.3			27	2,000	57		
28	20	0.6			28	3,450	98		
29	2,940	83			29	7,180	203		
30	19,400	549	31,000	878					
31	12,300	348	22,000	623					

plain. Beyond this function the structure does not provide and is not intended to provide downstream flood protection.''

An "as-built" plan view of the dissipator and levees is shown in Figure 6-2. Basically the dissipator consists of a channel 1 mi (1.6 km) long with rock revetment levees on each side. The width of the channel increases gradually from 230 ft (69 m) at the border to about 830 ft (249 m) at the downstream end. There is a low flow channel along the centerline. Surfacing of the bottom varies from concrete at the border, to grouted rock, to ungrouted rock, to earth at the terminus. Obviously the design objective was to slow the velocity by drastically increasing the roughness (n) factor and decreasing the hydraulic radius.

We could not find evidence of the reasoning behind the structure's orientation to the west. However, when we discovered that the structure was designed for a standard project flood Q of 135,000 ft^3/sec (4050 m^3/sec), the reason for the westerly orientation became obvious. A flow of this magnitude directed northwesterly could not, in any feasible manner, have been prevented from inflicting huge damage on the City of San Ysidro and on Interstate 5.

The velocity at the border, for a Q of 135,000 ft^3/sec (4050 m^3/sec), would be about 25 ft/sec (7.5 m/sec). Design of the structure was intended to reduce the velocity to about 7 ft/sec (2.1 m/sec) at Dairy Mart Road, which is still erosive but not exceedingly so. The structure was designed to induce a hydraulic jump at a point where the bottom of the channel was grouted. For the design Q the hydraulic jump would reduce the velocity from 25 to 17 ft/sec (7.5 to 5.1 m/sec).

Because the structure was designed for 135,000 ft^3/sec (4050 m^3/sec), it cannot possibly operate with the same relative effectiveness for a lesser Q. However, we found that, for the 1980 storm, it was at least as effective as having no project at all.

After the 1980 storms it was necessary to remove about 50,000 yd^3 (40,000 m^3) of silt that had deposited in the structure. It was our opinion that silt had settled toward the end of the second storm when the Q was much less than the peak 34,000 ft^3/sec (1020 m^3/sec). For any Q above about 10,000 ft^3/sec (300 m^3/sec), the velocities would have been too high for deposition. It appears that this silting problem will occur, to some extent, at the end of all heavy storms. A rather curious aspect of the situation that we could not fully explain was that in the latter stage of the storm most of the water was coming from the reservoir and the Mexican channel and should have been relatively clear. Why, then, was so much silt deposited in the structure? The answer

---

FIGURE 6–5

Tia Juana River discharges at the international boundary (from International Boundary and Water Commission).

appeared to be that the river is natural for about 7 mi (12 km) between Rodriguez Dam and the Mexican channel. Because the reservoir discharge was clear and "hungry," it obviously eroded a great deal of sediment from the bottoms and sides of the natural channel.

Rodriguez Dam

Our investigation of the dam and reservoir focused on the following questions:

1. Were the Mexican officials justified in believing that the dam was in jeopardy on February 20 when they opened all the floodgates? (Although the answer to this question may or may not have had legal relevance, we felt it was important to put the event into complete perspective. Our clients agreed.)
2. What were the volumes of water released, when were they released, and what effect did these discharges have on flood damage in the valley?

Many of the clues to the first question were found in a 1934 American Society of Civil Engineering paper written by Charles P. Williams, a Los Angeles engineer who was one of several United States engineers employed as consultants on the construction of the dam. The prime contractor, for both design and construction, was the Ambursen Dam Company, which had perfected a design for dams which bore its name.

The site selected was in a gorge, varying in width from about 100 ft (30 m) at streambed level to about 750 ft (225 m) at 130 ft (39 m) above the streambed. Flanked by a saddle, the gentle side slopes of the gorge rose to about 200 ft (60 m) above the streambed. The bedrock underlying the slopes was found to be badly weathered and contained several cleavage planes dipping toward the streambed. Nine borings were made in the streambed, and 52 test pits were excavated in the side slopes. The results and data were examined by a consulting geologist who opined that the rock in the streambed and on the western flank was suitable for a concrete dam foundation. However, he recommended an earth- and rock-fill dam as being more suitable for the eastern flank.

In the spring of 1928, despite concerns related to the disintegrated rock, design of the Ambursen-type concrete dam was completed and excavation for the foundations begun. Two adverse conditions were uncovered during excavation. One was that at 35 ft (10.5 m) below streambed level the bedrock was severely fractured, which was clearly not suitable for buttress foundations. In addition, a youthful geologic fault 20 ft (6 m) wide, not indicated on the surface, was discovered oriented in a direction nearly parallel to the streambed and along its easterly edge.

Now there was real cause for concern. Several prominent geologists from both sides of the border were consulted to reconsider the site itself. Topics examined included excessive reservoir leakage, finding adequate foundation material, cut-off wall design, and seismic considerations. The fault had been inactive for more than 100 yr but, considering the probable life span of the dam and the potentially disastrous results of dam failure, didn't the possibility of a 100-yr return period constitute too great a risk? (One hundred years is considered very youthful.)

For reasons not fully explained it was decided to use this site and to design to compensate for its weaknesses. It is possible that there were no other sites topographically acceptable that would provide the desired reservoir volume. Furthermore, indications were that the poor bedrock condition prevailed all along the streambed. To avoid foundation loads within the streambed, the dam was designed with massive reinforced ribs, spaced at 20-ft (6-m) centers, located above and tied to the concrete floor. The purpose of the ribs was to transfer the buttress loads by internal arch action to the sound rock on either side of the streambed. In addition to serving as a support for the buttresses, the ribs were intended to serve as horizontal bracing for the rock of the canyon walls where the cleavage planes occurred.

The final design of the dam was reviewed by the entire consulting board. A considerable number of design revisions were made by the group before construction began, and many more changes were made during construction to conform to field conditions. Although it appears that the design input was extremely competent and comprehensive, the quality control and construction supervision may not have attained that same level of excellence. For instance, it is rumored that payoffs by suppliers and subcontractors to General Rodriguez were so flagrant that a collector was permanently stationed at the site to receive the payola.

Construction was finally completed and the dam put into operation in 1932. Its main statistical features included

- Length of crest = 2000 ft   (600 m)
- Height of dam = 187 ft   (56 m)
- Highest maximum reservoir water surface = 180 ft   (54 m)
- Capacity of reservoir = 110,000 acre ft.   (13,530 ha m)
- Flood gates = nine 30 × 30 ft.   (9 × 9 m) structural steel gates on caterpillar bearings operated from tower
- Capacity of spillway = 150,000 ft^3/sec   (5400 m^3/sec)

After reviewing all the available data on the dam (including more recent geologic and geotechnical engineering reports) and discussions with International Boundary and Water Commission personnel, we concluded that the Mexican officials were justified in opening the gates when they did. That is not to say that it would not have been more prudent to have opened the floodgates earlier to spread the flow over a longer time frame and reduce the peak discharge. Information from the commission regarding reservoir levels during the two storms is summarized on the graph in Figure 6–6. The dramatic build-up that led to the opening of the floodgates on January 30 is significant in that it explains why there was so little time to react. On the other hand, the discharges during the second storm were the result of a more gradual water-level rise, making for easier control decisions. The experience of January 30 undoubtedly also played a big part in the earlier releases. Release should have begun sooner, but apparently orders to do so had to come from Mexico City. However, if the primary function of the dam was to impound water supply, what obligation did the Mexicans have to sacrifice their needs to provide better flood protection for the United States? Establishment of a "rule curve" to govern discharges would have been beneficial to both countries.

Sewer Rupture

An examination of the as-built plans for the interceptor sewer revealed the following pertinent information:

1. The pipe was 30 in. (76 cm) vitrified clay, a commonly used material for gravity sewers in Southern California.
2. The horizontal alignment of the pipe was such that it crossed under the eventual low flow channel, emanating from the dissipator structure at a point about 1500 ft (450 m) downstream from the end of the structure. Figure 6–2 illustrates this relationship.
3. The original cover over the pipe at the future crossing point was about 9 ft (2.7 m).

When the low flow channel was excavated, the cover was reduced to about 3 ft (0.9 m). Because the crossing point was so close to the end of the dissipator, the maximum velocity at the crossing point was probably in the range of 7–9 ft/sec (2–3 m/sec). This velocity, coupled with the highly erodible alluvial soil, would explain why the pipe was undermined and ruptured.

Either reinforced concrete pipe or concrete encasement at the crossing location might have survived the erosive action. The use of this would have required clairvoyance at the time the line was being designed. However, it appeared that the Corps of Engineers should have been aware of the existence of the pipe and should have incorporated its protection in the project design.

There seemed little doubt that the existing pipe would have withstood the flood if there had been no project and if the dissipator had been pointed northwesterly. It may have also survived if the Rodriguez Reservoir discharge had been more gradual.

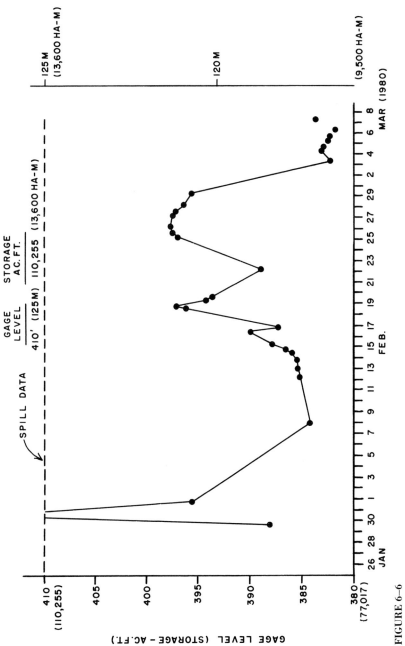

FIGURE 6-6

Rodriguez Reservoir levels (from International Boundary and Water Commission).

Analysis of Flood Damage

Even before receiving the results of the computer hydrology runs, we made a preliminary comparative analysis of the damage that actually occurred and that which would have occurred to valley properties as the result of identical storms and reservoir release conditions under four different scenarios:

1. With the entire project as it existed
2. With no project
3. With the Mexican channel in place but with no project on the United States side
4. With the entire project in place but with the dissipator structure pointed toward the northwest roughly along the alignment of the most recent low flow channel

A total of 27 land parcels plus 3 other entities—the interceptor sewer main, properties within the City of San Ysidro, and Interstate 5—were included in the analysis.

The scenario that assumed only the Mexican channel in place was obviously not realistic. Where it terminated at the border, the channel invert was about 7 ft (2 m) below grade. To outlet at grade would have caused a considerable widening of the Mexican channel. Velocities of 15–20 ft/sec (4.5–6 m/sec) at the terminus without some kind of energy dissipation would have caused tremendous erosion along the unpredictable path of the newly cut stream. Resulting damage could have been catastrophic to all the properties, including those within San Ysidro.

If the dissipator had been oriented toward the northwest (with corresponding changes in design of the levees), it appeared that three of the land parcels would have sustained considerably less damage, one would have received greater damage, the remaining parcels about the same level of damage, and the sewer main would probably not have ruptured. However, it appeared many of the San Ysidro homes and businesses would have been severely flooded and that Interstate 5 would have been threatened. This analysis was predicated on the calculated inability of the existing watercourse to handle more than 2000 ft³/sec (60 m³/sec) at its most critical section, allowing about 32,000 ft³/sec (960 m³/sec) to spill over. It appeared doubtful to us that it would have been feasible to contain the easterly spillover with earth levees. Further, we were convinced that a design Q of 135,000 ft³/sec (4050 m³/sec) would do considerably more damage if pointed northwesterly rather than westerly.

A comparison of actual damage to damage that would have resulted with no project seemed to be the most meaningful in terms of this lawsuit. Such comparison indicated that about three parcels lying easterly of Dairy Mart Road would have fared better with no project and the rest would have sustained about the same degree of damage. Later, the HEC-2 water-surface calculations confirmed our preliminary analysis.

Regarding the sewer-main rupture, it appeared to us the scouring of the low flow channel downstream from the dissipator structure was greater with the project than it would have been without. It was, therefore, our opinion that without the project the pipe might not have been ruptured. To argue that the pipe would have failed without the project would have implied that the original pipeline design was flawed.

Conclusions

Flood damage can be caused by one or any combination of three basic conditions:

1. Erosion (high velocity)
2. Deposition (low velocity)
3. High water

All three were present in the 1980 storms. Ordinarily these conditions occur over a relatively short period, corresponding to the peak flow time on a normal hydrograph curve. Because of the 2 days of controlled flow from Rodriguez Reservoir, the peak flow was artificially extended, thereby causing a much greater magnitude of erosion and deposition. Also, damage from high water was made worse as a result of standing for a longer duration. However, with several exceptions, we concluded that damage due to erosion, sedimentation, and high water was generally no greater than if there had been no project.

It was unfortunate that the energy dissipator was required. From a flood control standpoint, it would have proved much more cost-effective to extend the Mexican channel to the ocean and provide a settling basin for groundwater recharge. But, given the decision that was made, the dissipator was absolutely essential. There is no doubt in our minds as to whether the orientation of the dissipator was correct. If the flow had been directed to the northwest, along the alignment of the then existing low flow channel, total damage from the 1980 storms would have been much greater than it was. In the event of a design Q of 135,000 ft^3/sec (4050 m^3/sec), orientation of the dissipator toward the northwest would be calamitous.

Because of the ephemeral nature of the stream and the fact that its thread had meandered all over the floodplain during the last 150 yr, there was no reason to place greater emphasis on the most recent path in designing the dissipator and low flow channel.

If the dissipator had not been built, a new channel alignment would undoubtedly have been cut. The basic reason for this opinion is that the existing channel had a capacity of only 2000 ft³/sec (60 m³/sec) at its most critical section.

In our opinion the project functioned reasonably well in reducing velocities to preproject levels west of Dairy Mart Road and in spreading the flow back over the floodplain as it would have flowed without the project.

The rupture of the interceptor sewer could have been avoided if the Corps of Engineers had recognized the potential problems and had provided for its protection in the design of the low flow channel.

THE LITIGATION

The discovery and deposition phase of the litigation consumed many months. Each side employed a bevy of technical experts who were deposed at length. In addition, there was much additional deposition testimony from real estate appraisers, public officials (from United States agencies only), landowners, and eyewitnesses.

Deposition of the plaintiffs' experts revealed that they were going to present evidence to support their basic contention that there was greater flood damage with the project than there would have been without it. They were particularly critical of the orientation of the energy dissipator and of its effectiveness in dealing with a storm of lesser magnitude than that for which it was designed. However, at trial the plaintiffs' experts presented very little in the way of hydrologic or hydraulic calculations and relied mainly on their formidable backgrounds and reputations in expressing their opinions.

Trial was by judge in the State Superior Court of San Diego County. After weeks of testimony and examination of briefs presented by both sides, he rendered his decision. Essentially he found for the plaintiffs in the matter of the crop damage caused by the rupture of the sewer interceptor, and he found for the defendants in the matter of all other flood damage.

The judge's reasoning can best be presented by including his own letter in its entirety. Whether one agrees or disagrees with his final decisions, the letter is a good example of the ability of many jurists to absorb and sort out the technical complexities in a field foreign to their backgrounds and present their findings in a carefully reasoned and commonsense manner.

Gentlemen:

I have reviewed my notes of the testimony given, the exhibits admitted and counsels' briefs and am now ready to decide the issue of liability. I commend counsel for the thorough and professional manner in which the evidence was adduced and for the scholarly and literate briefs filed with the Court.

I will not in this letter attempt to make detailed findings of fact but will simply make note of impressions I deem significant to the decision. It is likely that counsel will request a more formal statement of decision and will furnish me with proposed statements which they feel are consonant with my conclusions. This letter will, however, go into the Court's file as part of the record in the case and will serve as a statement of decision if no more elaborate statement is requested.

I believe that for the purposes of this litigation the Tia Juana River Valley Flood Control Project (or "Project") must logically include the construction of the full concrete channel in Mexico as well as the dissipator system in the United States. The dams built earlier are not to be considered part of the "Project" but as part of the pre-existing watershed with which the "Project" was intended to cope.

There is arguably some difference in degree between the participation of the City and of the State in the Flood Control Project but both City and State supported its construction in general, provided financing for it, opposed its construction in a certain manner, appreciably influenced its design and approved the Project as it was ultimately designed and actually constructed. I conclude that both City and State were substantial participants in the Project and are jointly and severally liable in inverse condemnation for any damage which may have been legally caused by the Project.

It is not disputed that there was considerable damage to the plaintiff's properties by the flood waters. Crops, other vegetation, topsoil, structures, vehicles and other personal property were washed away or lost through burial. In addition, crops that remained were necessarily condemned because of contamination from sewage escaping from the rupture of the Tijuana Interceptor Sewer. Losses may also have been incurred through the disruption and delay of agricultural activities, either by reason of the time it took for flood waters to recede or because of the need to recondition the remaining topsoil or by reason of the lingering effects of deposits of sewage.

Was the project as deliberately designed and constructed a substantial factor in causing this described damage in the floods of January and February 1980? The evidence disclosed that the construction of the Mexican concrete channel materially increased the velocities of flood waters entering the United States. The concrete channel had as well the probable effect of reducing the sediment load which might otherwise have been accumulated in the passage of the waters through the floodplain in Tijuana, but the evidence indicated the amount of the reduction would have been so small as to be of little consequence.

On the United States' side of the border it appeared that the design and construction of the low flow channel within the dissipator and sedimentation area reduced the earthen cover above the interceptor sewer from nine feet to about three feet. It appeared that the dissipator system reduced the velocities of the 1980 flood waters to some degree but was not capable of reducing them to the level of what they would have been pre-Project by the point of their exit from the system at Dairy Mart Road. (Indeed, some of the witnesses for the City and State have conceded that the General Telephone parcel adjacent to and westerly of the realigned Dairy Mart Road did suffer increased damage by reason of the Project and that the velocities and scouring force of the flood waters were not reduced to pre-Project levels until leaving that parcel. As I understand it, however, there was an award of severance damages as to this parcel in an earlier eminent domain action and therefore General Telephone makes no claim in the instant case for damages to this property but only for damage to its other parcels.)

There was evidence also about the design and construction of the extension of the low flow channel after it left the dissipator system proper. The angle of intersection with the "natural" channel was criticized as being too acute. (I define "natural" channel as being the course of Tijuana River as it exited immediately prior to the floods of 1980, even though this course may have been in part "artificial" because of excavations of sand and gravel in borrow pits.) I believe it probable there was some "rebound effect" caused by the acutenesss of this angle.

The evidence established that the course of the Tijuana River has changed many times over the last one hundred years and that this would generally take place at times of flood. The river is described as "ephemeral" or "seasonal" or "braided," meandering widely in the Tia Juana River Valley on its journey to the sea. Except in years of heavy rainfall much of the flow from the watershed was retained by the Barrett Dam, the Moreno Dam and, in particular, the Rodriguez Dam, more recently constructed in Mexico in 1937. The annual rainfall at Moreno Dam for the 1979-1980 season was 41.4 inches, exceeding by an inch or more the record rainfall in 1916. The two preceding years had also been wet years; the dams were near capacity by the time of the storms of January and February 1980. The total rainfall from these storms, measured at Moreno Dam, was 30.3 inches. There were concentrations of rainfall at the end of January and beginning of February and especially from about February 13 to February 21 when over thirteen inches were recorded. It was apparently during these heavy concentrations that the Mexican Government determined to open the flood gates at Rodriguez Dam and did release great quantities of water for several days.

The experts called explained the stilling effect of the dams on water entering their reservoirs and the resultant loss of sediment load. The water released from the dams was therefore "clear" or "hungry" water. It was explained that the scouring capabilities of water depend primarily on its sediment load, including suspension and bed loads, and secondarily upon its velocity. This is denominated the "energy head." "Hungry" water scours more at a given velocity and scouring means not only loss of soil but meandering. Time also is a factor; it was explained that the extent of erosion was related to the duration of the flood. Finally, I learned that the course the flood waters took in 1980 when no longer contained or retained in the "natural" pre-1980 course was along the points of lowest elevation, for the most part southerly of the pre-1980 course.

Admittedly, velocities within the dissipator system, including the sedimentation area, were not in 1980 reduced by the system to pre-Project levels. I am of the view that these increased velocities coupled with the removal of most of the earthen protective cover from above the sewer in the construction of the low flow channel (in short, the Project) were a substantial factor in the rupture of the interceptor sewer and the contamination of the plaintiffs' crops and soil. The City and State have liability for loss of crops and any other form of loss which they may have suffered by reason of the contamination by sewage.

Was the Project a substantial factor in the damage to land and loss of crops and other property caused solely by the force of flood waters? Although I am most sympathetic with the plaintiffs and personally regret that a flood control project was not devised which would have protected them as well as other conflicting interests, I cannot in good conscience find it probable that the same damages and losses would not have occurred without the Project. I think to rule in the plaintiffs' favor on the point would require speculation.

The defendants' experts were of the opinion that the velocities of the flood waters were reduced to pre-Project levels either upon leaving the dissipator system or just westerly of that point, in or near General Telephone's most easterly parcel. Considering all circumstances mentioned heretofore and hereafter, the evidence supports that opinion. Although one might surmise that the "rebound effect" of the so-called angle of intersection might have set the southwesterly course of the new channel, it is more likely that the pre-existing elevations set this new course. There was, to be sure, unprecedented damage and loss, but there was also unprecedented rainfall concentrated in several days in January and February 1980, following wet years in 1978 and 1979. Added to the circumstances was the fact of unprecedented releases of large amounts of relatively sediment-free water from Rodriguez Dam. Counsel for the plaintiffs made a most able presentation on this issue but I believe it to be impossible of preponderant proof. I conclude that the Project was not a substantial factor in causing the loss to the plaintiffs by flood waters, either by changing the course of the waters or by the increase of the force thereof.

I will await word from counsel as to when they will be ready to proceed to the second phase of the trial; i.e., the determination of the nature and extent of the losses caused by contamination from sewage. I trust counsel will give some consideration to settlement negotiations now that a determination of the liability issue has been made. I must hold myself open for new assignments pending your report. I will try to avoid lengthy trials meanwhile.

THE POST MORTEM

This lawsuit was extremely complicated both technically and legally. There were so many technical variables that it was impossible to explain in precise mathematical terms the difference between preproject and postproject flood flows. Even the computer hydrology calculations had some limitations because of the constantly changing streambed elevations. Much of the technical analysis on both sides was based on judgment and experience.

The case constitutes a classic example of a major land-use and flood control policy decision made with little or no input concerning the threat and potential costs of liability. No implication is intended that the decision was not right, rather that in weighing the costs and benefits of the various alternative plans no cost contingencies were apparently included for the event that the adopted highly complex flood control system might not work exactly as planned. And when that happened litigation would not be far behind.

Luckily for them, the city and state were not held liable for direct flood damage. But does that mean that if a future flood occurs under slightly different circumstances (such as a higher Q, altered siltation characteristics, or state of maintenance of the dissipator structure) that there would not be a whole new round of lawsuits? And even if the defendants were to win again, the litigation itself is expensive.

One also wonders if the city and state included maintenance costs in comparing this system with a reinforced concrete channel. For instance, the levees require routine maintenance and the dissipator structure requires removal of large volumes of silt after every major storm.

Then there are two unthinkable events that would be devastating enough individually but that could happen simultaneously:

1. A runoff Q of 75,000 ft³/sec (2250 m³/sec), as occurred in 1916, or greater
2. Rodriguez Dam failure

It is mind-boggling to contemplate the damage and possible loss of life from such a combination. That is not to say that the dissipator would not handle up to 135,000 ft³/sec (4050 m³/sec) and perform as it was designed or that the situation would be worse than if there were no flood control project at all. But given the fait accompli of the Mexican channel and the complete lack of control over the operation and maintenance of Rodriguez Dam, wouldn't it have been more sensible for United States officials to have put a higher priority on flood protection for the whole valley?

The necessity of proper communication procedures in the event of sudden reservoir discharges is vividly illustrated in this case. Despite the additional problems posed by the event's international nature, it appears to us that there could have been closer coordination between the parties. It seems poor disaster planning and preparation that the belated decision to open the floodgates required the personal appearance of a federal official from Mexico City. Communications regarding floodgate openings to those that could be affected should be swift and foolproof. In this case, the Office of Emergency Services in San Diego should have been instantly warned. A recent disastrous flood on the Amazon River in Brazil was caused by the failure to communicate effectively to the operators of a downstream series of dams that the upstream dams were opening their floodgates.

As a related issue, this case also illustrates the general difficulties of dealing with the rights of downstream property owners in any agricultural floodplain when upstream property owners or government agencies perform grading or construct drainage improvements that alter downstream flow. It is basically a legal question, but one that engineers and geologists often face. It appears to us that case law is often confusing and contradictory. There need to be more definitive laws, ordinances, and guidelines for the people who plan upstream improvements, and government's obligation to the downstream property owners needs to be spelled out. This subject is explored in greater detail in Chapter 10.

7 FLOODING OUTSIDE A FLOODPLAIN: WAS IT REALLY A FLOOD?

THE EVENT

Woodland Hills is an affluent bedroom community located as far west as one can go in the San Fernando Valley and still be within the sprawling limits of the City of Los Angeles. The community is nestled in the northerly foothills of the Santa Monica Mountains, which form a steep chaparral-covered watershed.

As in most of Southern California, rainfall in the Woodland Hills area is confined almost exclusively to the winter months. However, its modest average rainfall of 15 in. (38 cm) is not indicative of the high-intensity storms that can occur in the rainy season. One such storm began with light rainfall during the early morning hours of February 13, 1980. The rain continued intermittently but with gradually increasing intensity for 4 days. Then at midafternoon on Saturday, February 16, the storm suddenly accelerated, resulting in peak runoff and culminating in the rapid inundation, 3–5 ft (1–1.5 m) deep, of over 30 residential properties and in severe damage to basement mechanical and electrical equipment at the famed Motion Picture and Television Relief Fund Home and Hospital.

The television evening news on February 16 showed, with graphic clarity, the magnitude of the inundation. In the eastbound lanes of the adjacent heavily traveled Ventura Freeway, vehicles were shown floating in what appeared to be 3 ft (1 m) of water. Curiously, traffic seemed to be proceeding without incident in the westbound lanes, where the water depth was only several inches.

Despite the abrupt rise of water within the residential streets to as high as 8 ft (2.4 m) and in the homes to windowsill height, there were no fatalities or serious injuries. One resident, in an attempt to reach his home on foot, was caught in the sudden surge and forced to swim to safety. Fortunately he was a strong swimmer. Many of the residents either hastily evacuated their homes or scrambled to their rooftops.

After cresting, the water surface ebbed nearly as quickly as it had risen,

Low-angle oblique photograph taken from helicopter at peak of flooding. Viewer is looking east along Ventura Freeway with the Dry Canyon twin box culvert approximately 650 ft (200 m) past the freeway sign. Photo shows the freeway median divider acting as a weir, causing the water to rise 2.5–3 ft (0.8–0.9 m) and to flow over the divider.

leaving a residue of mud, debris, and property damage. Early media reports contained only speculation as to the cause of the inundation. The unusually high intensity rainfall could not, in itself, account for why this neighborhood was more seriously affected than those adjacent.

THE INVESTIGATION

About 2 wk after the disaster we were contacted by a committee representing most of the affected property owners. At that time the owners had not yet hired an attorney. The committee employed us to conduct a preliminary investigation of the causes of the damages that had been sustained.

The primary purpose of this preliminary work was to help the committee decide whether to seek legal counsel. It was not part of our assignment to estimate the monetary amount of the damages. We suggested that this should be done without delay by consultants or contractors with specific appropriate qualifications.

View of the Ventura Freeway on-ramp at Valley Circle Drive. The deepest floodwater and box culvert is approximately 650 ft (200 m) to the right of the on-ramp sign.

Because the owners had been comparing notes and observations for 2 wk, we learned a great deal at our initial meeting with the committee, including

- The approximate limits of inundation
- The number and location of affected residences
- An account of the considerable damage at the Motion Picture and Television Relief Fund facilities
- The height of mud lines on some of the residences
- The approximate time of peak inundation
- A bizarre account of several automobiles being washed into the adjacent flood control channel at the height of the storm

Equipped with this input from the owners, we made our first field inspection. What we observed formed the basis of our entire investigation and preliminary report:

1. The drainage systems for the westerly Dry Canyon watershed which affected the residences (see Fig. 7–1) consisted of a concrete trapezoidal

FIGURE 7–1

Contributing watersheds to the 1980 Woodland Hills flood.

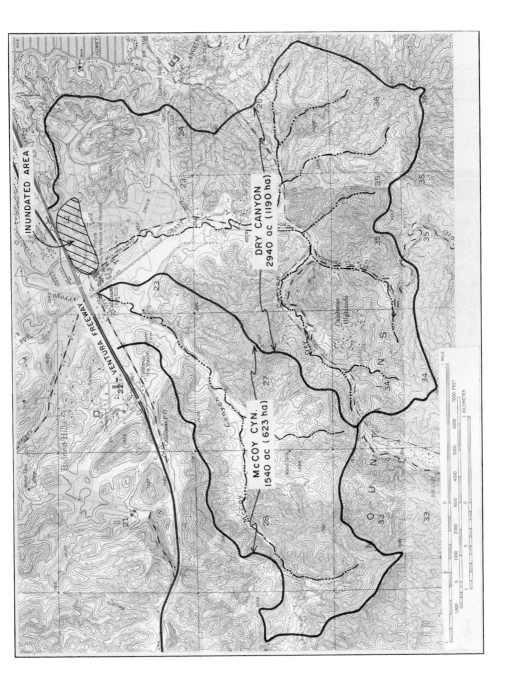

open channel, outletting to twin concrete box culverts under the freeway. It was readily apparent without measurement survey or calculations that these facilities were grossly undersized in relation to drainage structures both upstream and immediately downstream.

2. No underground storm drainage system existed within the residential subdivision comprising the lower portion of the Dry Canyon watershed. Considering the area of the watershed, this inadequacy was obvious.

3. The culverts under the freeway at the McCoy Canyon crossing appeared to have less hydraulic capacity than the upstream structures.

4. Neither canyon had any desilting or debris retention facilities.

5. Evidence of heavy erosion in Dry Creek Canyon consisted of uprooted trees, huge boulders, scoured slopes, etc.

6. Several large developments, involving extensive earthwork, were under construction in the upper reaches of the Dry Canyon watershed and were exposed to severe erosion. It was evident that temporary erosion control measures, which are required during the rainy season in Los Angeles County, had been either inadequate or nonexistent.

7. The freeway longitudinal grade was in a sag vertical curve at the Dry Canyon crossing. The freeway median strip contained what appeared to be a recently installed solid concrete barrier about 3 ft (0.9 m) high that obviously acted as a weir and a dam.

Predicated largely on this initial field inspection, we determined a program for our preliminary investigation:

- To roughly estimate the maximum flow (Q) produced by the storm in the two watersheds
  (a) Unbulked (water only)
  (b) Bulked (with sediment and debris)
- To estimate the magnitude of the storm in terms of return frequency probability
- To correlate the theoretic time of maximum flow with information supplied by the property owners
- To calculate the approximate capacity of the pertinent drainage structures
- To assess the effect of blockage or partial blockage of the freeway culverts by debris
- To assess the effect of the freeway median barrier on the water-surface levels
- To determine how and why vehicles had been washed into the channel, what had happened to them, and what effect their presence had on water levels
- To test the sediment deposited in the downstream portion of the watershed to ascertain if and where it had been eroded upstream
- To contour flood elevations on the basis of mud lines

One of the first things to do in any forensic investigation is prepare a chronologic list of all conceivably significant events in connection with the disaster. Included are the actual physical happenings, reports, construction contract documents, letters, memoranda, legal proceedings to date, photographs, aerial photographs, and any other pertinent items. The list serves a dual purpose. First, it aids in putting the whole event in an historical context. Second, it can be used as a handy reference during questioning at deposition and trial. One of the big problems encountered by experts (especially the older ones) is memory loss. One can be involved in a case, on an intermittent schedule, for as long as 5 yr. This means that a tool, like a chronology, can be very helpful for a quick review of the case.

Another helpful initial effort is to use the data collected to recreate the disaster step by step. Eyewitness reports and questionnaires, along with media accounts and film, can be used effectively for this purpose. Often there can be great discrepancies among eyewitnesses as to time, sequence, and magnitude of the component events. However, it is usually possible to resolve such differences by arriving at a consensus and comparing indirect sources of information such as emergency telephone communications and police and fire reports. In the case of flood disasters, automatic rainfall and stream-flow gages, if available, are extremely valuable instruments for recreation.

We found the chronology of construction pertinent to this event to be as follows: The Ventura Freeway, including its drainage structures, was constructed in 1957 by the California State Division of Highways (now called CalTrans). The state farmed out the design responsibility for this segment of the freeway to the City of Los Angeles Department of Public Works. Design criteria were those of the state. The state contracted for construction and, upon completion, assumed responsibility for maintenance and operation functions.

Prior to construction of the freeway, Dry Canyon had basically been a natural watercourse outletting to a corrugated metal pipe culvert under the old Ventura Boulevard.

Storm drainage improvements in the freeway contract consisted of realignment of the watercourse; construction of twin concrete box culverts 10 ft (3 m) wide by 8 ft (2.4 m) high under the freeway and frontage roads; and grading of earth channels between the freeway and Leonora Street, from Leonora downstream about 500 ft (150 m) and from Avenue San Luis Rey upstream about 150 ft (45 m).

The residential subdivision, which bore the brunt of the damage, also was developed in 1957. In conjunction with the subdivision, the private developer designed and constructed the following drainage improvements:

1. A trapezoidal channel with a concrete bottom 12 ft (3.6 m) wide and 1.5 : 1 Gunite side walls 6 ft (1.8 m) high, with a design water surface

4.5 ft (1.35 m) high. The channel extended upstream from the freeway culvert entrance about 350 ft (105 m).

2. A cutoff wall, concrete piers, chain link fence, and Gunite spillway at the westerly terminus of Bigler Street

These improvements were designed by private engineers, approved by the city and the Los Angeles County Flood Control District, constructed by the developer, and then dedicated to the district for maintenance.

In 1960 another subdivision was added, resulting in an upstream 1400-ft (420-m) extension of the channel.

In 1963 the city designed and in 1970 constructed a concrete box culvert 19 ft (5.7 m) wide by 14 ft (4.2 m) high at the Dry Creek–Mulholland Drive crossing.

In 1972 the flood control district constructed a rectangular concrete channel 19 ft (5.7 m) wide by 14 ft (4.2 m) high, beginning at the outlet of the freeway culverts at Leonora and discharging into the Calabasas Creek Channel, about 1500 ft (450 m) downstream.

In the late 1970s another system of storm drains was designed and built by the district. This system was located about 0.5 mi (0.8 km) upstream from the Mulholland Drive culvert.

In about 1977 CalTrans replaced the chain link fence barrier in the freeway median strip with a solid precast concrete barrier.

Storm drain construction in the McCoy Canyon watershed consisted mainly of replacing culverts under old roadways or installing new culverts under new roadways as development progressed. The McCoy Canyon crossing at Park Sorrento was handled by two 10 × 10-ft (3 × 3-m) concrete box culverts of recent vintage. This was in sharp contrast to the old downstream culvert at Calabasas Road, a single concrete arch 4 ft (1.2 m) wide by 6 ft (1.8 m) high. At what is now the freeway crossing, there was an existing concrete box culvert 8 ft (2.4 m) high by 9 ft (2.7 m) wide under the old Ventura Boulevard. This culvert was left in place for the freeway crossing with no additions.

Having now assembled the preliminary information from the property owners, public agencies, and site inspection, we began our analysis.

Watershed Delineation

Watershed boundaries were plotted on topographic and tract maps obtained from the city and county and then planimetered to estimate areas contributing runoff to the freeway culvert entrances.

Dry Canyon watershed 2940 acres   (1176 ha)
McCoy Canyon watershed 1540 acres   (616 ha)

We then plotted on the same maps all the pertinent drainage structures for which we had obtained plans or which we noted in the field. Figure 7–2 shows the major features that formed the basis for our analyses.

Magnitude of Rainfall and Runoff

Rainfall records were obtained from the flood control district for all its automatic and standard gages in the general area. After studying the records we chose the Encino Reservoir automatic gage, located about 5 mi (8 km) east

FIGURE 7–2

Site map showing pertinent drainage facilities.

of the site, as probably being more comparable to actual site conditions than any of the others or any combination. Reasons for the choice included proximity, isohyetal compatibility, and that it was an automatic gage that recorded rainfall in 15-min increments.

Table V depicts the results of our analysis of the Encino gage records for February 16, 1980.

Using flood control district criteria for velocity-discharge-slope relationships in both mountain and natural valley channels, we estimated the time of concentration for each watershed to be about 30 min. Later we found corroborating documents from the district indicating that it had calculated about 25 min for each watershed. The district also published average rainfall intensity duration curves for various recurrence or return intervals. Figures shown in the last two columns of Table V were derived from these curves. The table indicates that for a 30-min duration, a 25-yr storm has an average intensity of 1.75 in./hr (4.4 cm/hr) compared with an actual intensity of 1.80 (4.6 cm) for the subject storm. This comparison was the basis for our estimate of a 25-yr return. Had the time of concentration been greater—in the 1–5-hr range—the return intervals would have been in excess of 50 yr.

We found that there were no stream gages in either Dry Creek or McCoy Canyon to measure actual flow. However, it did not appear necessary, at least at this preliminary stage, to spend a great deal of our clients' money for a comprehensive computer hydrology study to determine the exact peak flow

TABLE V
Rainfall Intensities

| Peak period | Time peak period began | Rainfall | | Equivalent hourly rate | | Approximate rate for recurrence interval | | | |
| | | | | | | 25 yr | | 50 yr | |
		(in.)	(cm)	(in.)	(cm)	(in.)	(cm)	(in.)	(cm)
15 min	4:00 PM	0.5	1.3	2.0	5.1	2.4	6.1	2.8	7.
30 min	3:45 PM	0.9	2.3	1.8	4.6	1.75	4.4	2.1	5.
1 hr	3:15 PM	1.6	4.1	1.6	4.1	1.2	3.0	1.6	4.
2 hr	3:15 PM	2.6	6.6	1.3	3.3	1.0	2.5	1.2	3.
3 hr	2:30 PM	3.6	9.1	1.2	3.0	0.85	2.1	1.05	2.
4 hr	1:30 PM	4.7	11.9	1.17	3.0	0.8	2.0	1.0	2.
5 hr	1:00 PM	5.2	13.2	1.04	2.6	0.75	1.9	0.9	2.
12 hr	5:45 AM	7.9	20.1	0.66	1.7	—		—	
24 hr	7:30 PM (2/15)	8.3	21.1	0.35	0.9	—		9.5	24.
Complete duration of storm (2/13–2/17)		15.4		0.21		—		—	

for this storm. We thought it could be roughly estimated by means of the Rational formula ($Q = ACI$) despite it not being considered a reliable approach for watersheds greater than 100 acre (40 ha).

Also, since the rainfall appeared to be in the 25-yr return range, there seemed no reason (such as a recent brush fire in the watershed), the return Q was not in same probability range. Also, for Qs of this return interval and greater the district had already performed detailed hydrologic studies. Our rough estimates of the Q at the entrances to the freeway culverts for this particular storm were 4400 ft^3/sec (132 m^3/sec) for Dry Canyon and 2300 ft^3/sec (69 m^3/sec) for McCoy Canyon.

Both these figures included a bulking factor of 20% to take into account the estimated volume of sediment and debris eroded in the undeveloped portions of the watershed as well as those areas being graded at the time of the storm and that lacked adequate erosion control. The design Qs calculated by the district were as follows:

Return interval (yr)	Dry Canyon		McCoy Canyon	
	(ft^3/sec)	(m^3/sec)	(ft^3/sec)	(m^3/sec)
25	3800	114	2200	66
50	4880	146	2700	81

These figures were based on future full development of the watersheds and therefore contained no bulking factor. At the time the watersheds were only about 30% developed. The remaining undeveloped 70% consisted of moderate to steep hillside terrain, which, according to district criteria, had a peak bulking rate of about 1.5. This caused us to wonder whether the determination of a design in this type of terrain (both topographically and geologically) should not call for two hydrologic calculations—one based on the conditions as is and one with ultimate development—with the higher of the two figures prevailing.

We speculated that the fact that bulking was not considered may have been the reason why there were no desilting basins or trash interceptors. If true it meant there would be an initially considerable but constantly diminishing risk during the development period.

Investigations of old city hydrology plans and documents indicated that in 1957 the city used state criteria in calculating the peak flows for design of the freeway culverts. This criteria specified a 25-yr storm, no bulking factor, and no consideration of future development. It resulted in Qs of 3260 ft^3/sec (98 m^3/sec) for Dry Canyon and 1860 ft^3/sec (56 m^3/sec) for McCoy Canyon.

Our conclusions relative to the various peak flow calculations were as follows:

• There had been no adoption of uniform hydrologic criteria among the public agencies despite the fact that their facilities were connected within the same watercourse.

- Los Angeles County Flood Control District criteria and policy seemed much more logical and sophisticated than that of the state.
- It would not be necessary for us to perform a computer hydrology analysis for the subject storm since we already had enough data to estimate adequacy of the culverts and channel in terms of a 25-yr storm and in terms of the existing structures both upstream and downstream.

Drainage Structure Capacities

Using the "as-built" drawings, we calculated the capacities of the critical drainage structures as follows:

1. Dry Canyon freeway culverts   1920 ft³/sec   (58 m³/sec)
2. Dry Canyon trapezoidal channel 1860 ft³/sec   (56 m³/sec)
3. McCoy Canyon freeway culvert   780 ft³/sec   (23 m³/sec)

By contrast, the flood control district's rectangular channel downstream from Leonora had a capacity of 4880 ft³/sec (146 m³/sec), while the city's upstream culvert under Mulholland Drive had a capacity of 4000 ft³/sec (120 m³/sec). Calculations for the Dry Canyon freeway culverts were based on an entrance water-surface level at the top of the headwall. It is interesting to note the relationship between capacity and elevations for these culverts (see Fig. 7–3). The abrupt decrease in capacity that occurred when the culverts were flowing nearly full (as a channel) as opposed to when flowing full under pressure became a subject for discussion as explained later in the chapter.

The channel capacity calculations were predicated on 1.5 ft. (0.5 m) of freeboard. This is a safety factor standard based on a supercritical velocity of 22 ft/sec (6.6 m/sec) on a curving alignment, and is intended to take into account water-surface rise due to turbulence and superelevation in addition to providing space for floating debris such as tree trunks.

Table VI summarizes capacities, design Qs, and deficiencies for the three critical drainage structures.

Fitting the Pieces Together

We now had enough data to begin to establish a probable scenario for the event.

A survey of high-water marks and mud lines on the houses indicated that the water level had attained a maximum elevation of about 923 ft (277 m). The height of the freeway median barrier was measured at 2.67 ft (0.8 m). Adding

TABLE VI
Summary of Culvert and Channel Deficiencies

Design agency	Year of design	Q (ft³/sec)	Q (m³/sec)	Deficiency (ft³/sec)	Deficiency (m³/sec)	Percent deficient
		Dry Canyon Channel (capacity = 1860 ft³/sec [56 m³/sec])[a]				
City of LA	1956 (25 yr)	3030	91	1170	35	39
LA County Flood Control District	1964 (50 yr)	4880	146	2420	73	57
LA County Flood Control District	1980 (this storm)	4800	144	2940	88	61
		Dry Canyon culverts under freeway (capacity = 1920 ft³/sec [58 m³/sec])[b]				
City of LA	1956 (25 yr)	3256	98	1336	40	41
LA County Flood Control District	1964 (50 yr)	4880	146	2960	89	61
LA County Flood Control District	1980 (this storm)	4800	144	2880	86	60
		McCoy Canyon culverts under freeway (capacity = 780 ft³/sec [23 m³/sec][c]				
City of LA	1956 (25 yr)	1860	56	1080	32	58
LA County Flood Control District	1964 (50 yr)	2700	81	1920	58	71
LA County Flood Control District	1980 (this storm)	2200	66	1420	43	65

[a] Based on 1.5 ft (0.45 m) freeboard per design drawings.
[b] Based on water surface at top of headwall.
[c] Based on water 3 ft (0.9 m) over soffit of culvert.

this height to the freeway centerline elevation at the lowest point of the sag vertical curve established the minimum elevation of the top of the freeway barrier at about 922.5 ft (276.7 m). This then confirmed the surface elevation of the back-up pond and permitted us to plot its boundaries on a topographic map.

A Bigler Street resident provided vital information regarding the three automobiles washed into the channel. At about 3:45 PM, his neighbor's car was the first to go through the chain link fence and be swept into the muddy debris-laden torrent. At this point the channel was already overflowing, so its walls could not be seen. It was not known whether or for how long this first car may have been hung up at the entrance to the culverts before passing through. But pass through it did, because its remains were found several days later in the Sepulveda Basin of the Los Angeles River, about 8 mi (12.8 km)

Station wagon that became impaled on the entrance to the freeway culverts being removed the day after the inundation occurred.

downstream. Within seconds, the second vehicle, belonging to the witness's maid, floated over the flattened fence and into the channel. Again there was no evidence as to how long it may have acted as an obstruction at the culvert entrance. This car was never found. The third vehicle, belonging to the wife of the witness, a large station wagon, never made it through the culverts. Probably because of its size, it became impaled on the entrance and was not recovered until the following morning by county maintenance workers.

Fortunately for our investigation, one of the homeowners who found refuge on his roof had the presence of mind to bring his home movie camera. His adept camera work provided clear evidence of the sequence of events that caused the water transport of the cars. The film revealed that the cars began to float when the water reached a depth of about 18 in. (0.5 m). Then, as the water continued to rise, they began to move slowly—not downstream but upstream!

This constituted proof that the force of the water back-up was initially greater than the force of the downstream flow in Bigler Street (a fact that would later prove valuable). The film showed that when the water had risen to nearly full crest, the cars reversed direction and began to move downstream. The first car to approach the fence appeared not to break through the chain link mesh but to push the fence posts over from the tops of the footings. The apparent ease with which the fence was flattened prompted us to obtain copies of the original construction plans from the city. In comparing these plans with actual field conditions as first observed, we found that a continuous reinforced concrete pier wall had been omitted in construction, leaving only individual shallow footings at the fence posts. We employed a structural engineer to calculate the theoretic difference in horizontal forces causing failure between the fence as designed and the fence as actually built. The results indicated that the fence as designed would have been about 66% stronger in terms of lateral impact than the actual constructed fence. Therefore, the fence as designed would have stood a much greater chance of restraining the vehicles.

The sequence and causal relationship of events leading to the damage could then be summarized as follows:

1. The storm produced a peak flow of about 3800 ft^3/sec (114 m^3/sec) in Dry Canyon and 2200 ft^3/sec (66 m^3/sec) in McCoy Canyon.
2. Culverts under the freeway at these locations had capacities of only 1920 ft^3/sec (58 m^3/sec) and 780 ft^3/sec (23 m^3/sec), respectively.
3. The trapezoidal concrete channel immediately upstream of the Dry Canyon freeway culverts had a capacity of only 1860 ft^3/sec (56 m^3/sec).
4. When the capacities of these culverts were reached, water began to back up and to overflow onto the freeway.
5. There were no debris basins or trash barriers in either watercourse to inhibit the flow of mud and debris from areas where construction erosion control facilities were inadequate.
6. As the debris-laden water continued to rise, the freeway median barrier acted both as a weir and a dam, preventing downstream escape and elevating the water flow by approximately 2.7 ft (0.8 m).
7. When the water depth reached about 18 in. (0.5 m) at the cul-de-sac in Bigler Street, three vehicles began to float.
8. The heavy surface discharge onto Bigler Street from the developed upstream watershed caused the cars to wash through a chain link fence (which did not have the resistance to impact that it would have had if it had been properly built) and down to the freeway culverts.
9. At least one of the vehicles was a substantial source of blockage during peak flow, thereby further decreasing culvert capacity and causing the water to rise faster.
10. Overflow and back-up from the inadequate McCoy Canyon culvert joined that of Dry Canyon to form one large pond.
11. Relief was only attained when the water rose to the top of the freeway median barrier. By that time considerable damage had been done.

THE LITIGATION

After considerable investigation, meetings with the homeowners, and review of our preliminary report, the plaintiffs' attorneys filed suit against the home-owners' insurance companies, CalTrans, Los Angeles County, and the City of Los Angeles. Suits against all defendants included claims for physical damage as well as bad faith and punitive damages. Suits against the public entities also included claims for inverse condemnation, the aforementioned legal concept wherein property owners are entitled to compensation if they can prove their property has been damaged ("taken") as a result of a government project or action.

The discovery process included many months of information exchanged in the form of interrogatories, document subpoenas, and depositions. Because of the multitude of insurance company defendants, the mechanics of trial scheduling became quite complicated. Basically the Superior Court judge decided to consolidate the defendants and trifurcate (divide into three phases) the trials. The first phase would involve the public entities, the second phase the insurance companies, and the third phase the matter of bad faith and punitive damages. Ultimately the first and second phases were combined except for one insurance company that had its own individual trial in a federal court.

All this legal maneuvering did not affect us except that we had to keep our own schedules flexible. In essence, since the first trial would presumably resolve issues of causation, it would be where most of our testimony would be heard.

One of the facts disclosed by discovery was that the flood control district, in a series of letters to CalTrans over many years, had proposed a joint project to upgrade both the freeway culverts and the upstream channel. CalTrans steadfastly refused the offer, preferring to wait until such time as it widened the freeway. This appeared to constitute an obvious case of not correcting a known hazard.

The insurance companies had, in general, refused payments of home-owners' claims on the grounds that their standard policies did not cover damages due to flooding. This position created an interesting semantic question. All the technical definitions of flooding we could find in our references indicated that flooding is a condition that occurs when watercourses (either natural rivers and streams or improved channels) overflow their normal boundaries. In this case the inundation was the result of the *back-up* of water from artificial constrictions and barriers rather than from overflow of the canyons. Although the legal definition of the word flooding never became an issue in the trial, its implication may have been a factor in the insurance companies' eventually settling. It was our impression that the insurers did not want to risk an adverse court decision that might set a precedent for cases not fitting a strictly defined concept of flooding.

Another dilemma facing the insurers' attorneys was how closely to ally themselves with the other defendants, which were all government entities. Obviously their first choice was to win the cases along with the other defendants. However, in the event the plaintiffs won, they wanted to be in a subrogated position to recover by suing their codefendants for negligence. As a result they did not appear to attack the plaintiffs' arguments and evidence of negligence wholeheartedly.

Principal allegations against the state, county, and city included the design and maintenance of the freeway culverts and flood control channel, construction of the chain link fence, failure to provide drainage openings in the freeway median barrier, and failure to provide desilting or trash interceptor facilities.

Because of recent applicable court decisions, the plaintiffs' attorney did not consider it necessary to prove negligence to the insurance companies, rather that the damage was caused by the action of third parties. In other words the key question was, Would damage to the plaintiffs' properties have occurred if there had been no freeway, no culverts, and no flood control channel? Our studies indicated that damage under these circumstances would have been minimal since only several of the lots on the easterly fringe would have been within the natural floodplain.

Our depositions in the case were neither prolonged nor particularly eventful. Most of the questioning was done by attorneys for the insurance companies, who tended to concentrate on our original preliminary report.

Depositions of the defendants' experts made it clear the direction in which the defense attorneys were headed:

- To show that the magnitude of the storm was close to a 75-yr return frequency rather than 25 yr (if successful, this would tend to put the event in more of an act of God classification)
- To make the freeway culverts appear to be less inadequate in capacity than we had calculated
- To show that the channel had greater capacity than we had calculated
- To show, by means of accident statistics, that providing for drainage through the freeway median barrier would result in decreased structural strength, thereby enhancing the possibility of head-on collisions
- To show that the three automobiles were washed into the channel by runoff from the upstream watershed and did not begin to float as the result of back-up from the channel
- To show that the construction deficiency in the chain link fence footings was not a factor in the failure of the fence to restrain the cars from washing into the channel
- To argue, in general terms, that the disaster was caused by a combination of circumstances and events that prudent engineers could not reasonably have been expected to foresee

The deposition testimony by the defendants' experts sent us scurrying to check our investigations, analysis, and calculations. The results:

- We concluded that our estimate of a 25-yr return frequency storm was essentially correct. We found a basic error on the part of one of the defendants' experts in transferring a key rain gage reading to be the primary reason for the discrepancy.
- We found that some of the defendants' experts had calculated the capacity of the culvert based on it flowing not quite full (as an open channel), yielding a 20% higher capacity than if the culverts were flowing full with entrance head to the elevation of the roadway surface. This approach, in effect, ignored entrance and soffit friction loss. It was not realistic since the maximum capacity would only occur momentarily, that is, once when the water rose and once when it fell. Since culverts are not designed on this basis, it is illogical that their existing capacities be investigated on this basis. We felt confident about our methodology. See Figure 7–3.
- One of the defendants' experts, in calculating the capacity of the channel, had measured cross-sectional areas from certain of the as-built drawings that were in conflict with those we used and that yielded a greater cross-sectional area than ours. Because the difference was fairly significant we felt compelled to have the channel surveyed. The survey revealed a slightly smaller cross section than we had used, thus making our results conservative. The same expert also ignored freeboard in his calculations, thereby greatly increasing the channel's capacity. We had incorporated a 1.5-ft (0.5-m) freeboard in our calculations, which conformed to flood control district design criteria for channels with velocities greater than 15 ft/sec (4.5 m/sec). The fact that part of the channel alignment was curved, plus the potential debris problem, made inclusion of reasonable freeboard even more essential. Again, why should proper design criteria differ from that for investigation of adequacy?
- In examining the CalTrans design manual, we found that one of the standard median barriers was a metal beam type. Though slightly more expensive, this barrier was structurally equivalent to the solid precast concrete barrier that had been installed. We then calculated its ability to pass the excess volume that the culverts were not able to handle. While not fully capable of passing the entire excess flow, this barrier, had it been in place, would have reduced the high-water level to the point where negligible damage to the structures would have occurred.
- The questions of how and when the cars began to float was, of course, simple for us to verify on the basis of the homeowner's movie film. When the force of the downstream current exceeded that of the back-up, the cars were washed westerly through the fence and into the channel. Thus it seemed obvious that the cars would not have floated as a result of only

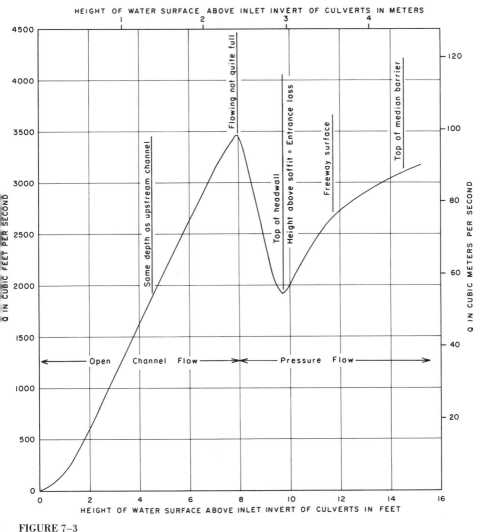

FIGURE 7–3

Relationship of culvert discharge to water level.

the overland flow on Bigler Street. To prove this we calculated the depth of flow in Bigler Street based on its upstream watershed only. Maximum depth was about 10 in. (25 cm). Research indicated that generally cars do not begin to float until the water depth becomes about 18 in. (46 cm). The movie film confirmed this research.

• The movie film was also instrumental in analyzing the role of the chain link fence as previously described. One of the defendants hired an expert who opined that the fence posts (which were standard diameter and gage) had failed prior to the cars washing through. His calculations were based on the assumption that the fabric openings had been clogged with debris and that the water level was higher on the upstream side, thus forcing the fence to act as a partial retaining wall. The movies did not bear out his theory. They showed that the fence was intact at the time of initial impact and that the fence posts were pushed over, with failure obviously occurring at the footings.

- The inevitable argument by the defendants that this disaster was unpredictable because it was caused by a chain of highly improbable events and therefore was an act of God gave us considerable cause for reflection. After all, we were not advocates. Our job was to gather the facts, analyze them, and report our findings to the attorney. It was then his job to use them as he saw fit within the framework of his legal strategy. However, from a scientific and professional view we felt an obligation to help the attorney refute the act of God argument. True, it took a series of events to cause this disaster. But was the disaster predictable with reasonable scientific probability? Broken down, event by event, we thought it was.

First, the storm was in the 25-yr return frequency range. Most major storm drainage facilities are now designed for at least 50-yr and usually 100-yr discharges. Certainly unpredictability is not evident there. The fact that the state had recently installed a solid concrete traffic barrier in the freeway median, even at the sag curve, seemed logical enough. Similar installations were being made throughout the state. However, in this case, the state had been informed in writing by the flood control district many times over a 10-yr period that the culverts were inadequate. In all probability this fact was never communicated to the people who were planning the median barriers or, if communicated, was ignored. In fact, nobody probably even thought of the possible consequences. This action might compassionately be termed an oversight, but certainly not an act of God.

The presence of the parked cars near the Bigler Street cul-de-sac at the time of the storm was not unusual or unpredictable. In our visits to the site we frequently observed the presence of parked cars in this area.

Neither was the inability of the fence to restrain the cars from washing into the channel unpredictable. The fence was designed and constructed basically for security purposes—to keep people out of the channel right-of-way—not to act as a debris fence. Even if fence-post footing construction had been in compliance with the plans, there was still no assurance that the cars would have been restrained.

After analyzing each of these events individually, we concluded that the disaster was predictable within reasonable scientific probability standards. We so informed the attorney. We felt so strongly about this point that we urged the homeowners to exert whatever political pressure they could to force CalTrans not to delay action until the freeway widening project but to replace the median barrier immediately. Even after the case was settled, it took CalTrans about 3 yr to take this action.

In general the defendants' experts spent much time and money on their investigations. Two of them prepared HEC-1, HEC-2, and other computer hydrology studies to determine peak flows, water-surface elevations, and flood storage. The results of these studies pretty well confirmed what actually happened, at least up to the time the automobiles became a factor in the equation.

The main trial involving the largest number of insurance carrier defendants was never completed. During early presentations of testimony and evidence by the plaintiffs' attorneys, the insurance companies and the government entities began to settle individually. By the time expert testimony was half completed, they had all settled.

THE POST MORTEM

In the wake of the disaster two steps were taken to reduce the impact of a potential repetition. In 1983 the Los Angeles County Flood Control District contracted for the installation of an underground storm drainage system in the developed westerly portion of the Dry Creek watershed. The system minimizes the flow of surface water in the streets discharging into Bigler Street, thereby decreasing the probability of vehicles being washed into the Dry Creek channel.

However, the system did nothing to prevent back-up from the inadequately sized culverts under the freeway in the event of blockage from debris transported down the channel from the easterly undeveloped portion of the watershed. As both watersheds continue to be rapidly developed, the runoff increases accordingly, making the culvert increasingly inadequate. In addition, the new storm drainage system decreased the time of concentration for the westerly watershed, thereby increasing the maximum flow.

The second mitigating action, taken by CalTrans about 5 yr after the disaster, was to replace a segment of the freeway concrete median barrier, at the longitudinal sag point, with a metal beam barrier. This permitted most flow in excess of the culvert's capacity to escape northerly across the freeway with minimum ponding. Why this improvement was delayed so long after the case had been settled defies understanding.

The general philosophy of the public agencies, particularly CalTrans, toward mitigation of the problem seemed to be:

- The event had been caused by an unlikely succession of circumstances whose probability of repetition was exceedingly low. This belief gave mitigation measures a low priority.

- The problem would be solved when CalTrans widened the freeway and installed additional culverts. The widening project was finally completed in 1990, 10 yr after the disaster and 25 yr after the flood control district had initiated correspondence with CalTrans regarding a joint improvement project, but without additional culverts! We have not attempted to discover why the culverts were not included in the contract but suspect that they were considered not to be cost-effective.

This philosophy provokes the following questions:

1. Should the residents have been exposed to risk for such a long period when a relatively inexpensive partial mitigation measure (the change in the median barrier) was available?
2. Should there not be a simple legal procedure whereby public entities could provide mitigation during litigation without prejudicing their legal position? The need for such a procedure is particularly great when one considers it often takes 5 yr for these cases to come to trial, plus as long as a year for the trial itself, plus possible appeal time.
3. Assuming the new underground storm drainage system will lessen the possibility of automobiles being washed into the channel, what is the possibility that other large objects such as tree trunks and boulders will cascade down the channel from the undeveloped portions of the watershed? Culvert blockages usually occur at or near the time of peak flow. This is because maximum transport velocity coincides with peak flow.
4. What effect would a major brush fire within the watersheds (so common in Southern California), followed by a major storm, have on both the peak flow and the quantity and size of rocks and debris subject to erosion? According to the county's hydrology manual, peak flow would be increased about 60% by a fire–flood sequence in these watersheds.

Some of the lessons we think should have been learned by this disaster and the litigation that followed are:

- The design of freeway culverts and other drainage structures at vulnerable locations should be in accordance with worst-scenario criteria regarding both storm return frequency and runoff coefficients *based on future possible development*. Generally the cost of larger drainage structures is insignificant in comparison to the cost of total freeway construction and land acquisition costs. Further, the initial cost of adding an increment or two in the size of a culvert or drainage structure is miniscule compared with the cost of enlarging the facility after the freeway is built and operating. This case is a classic example.
- Traditionally in many public agencies the decisions regarding adoption of design criteria have been based on cost–benefit analyses, taking into

account only initial construction. The cost side of the equation must include the present value of maintenance costs over the life span of the facility. And, of paramount importance, there must be an attempt to evaluate and incorporate the potential costs of litigation, including awarding of damages, should someday the design be judged inappropriate or inadequate, the construction flawed, or the maintenance faulty.

- Drainage structures should not be underdesigned simply because existing structures, either upstream or downstream, are substandard. Though the structure, properly designed, will not function at full efficiency until the rest of the system is upgraded, it invariably proves to be false economy not to design for the ultimate system. This particular case provides a good example. The trapezoidal concrete channel was designed by a private engineer working for the developer. The City of Los Angeles approved the design for the upstream section in 1964 based on a design flow that obviously coincided with the actual capacity of the downstream freeway culverts. Someday the channel, which is still in good physical condition, will have to be completely rebuilt if and when the number of freeway culverts is increased, but this time at the expense of the taxpayer, not the developer.

- Finally, there is the perfectly logical argument by government entities that their budgets will not permit them to upgrade or enlarge all their old drainage facilities that are known to be inadequate in terms of present conditions and criteria. No one questions this. However, it is advantageous for public entities to analyze and prioritize these deficiencies to take into account not only potential physical damage but the threat of litigation. If the presence of an inadequate culvert poses no greater hazard to surrounding properties than if there were no culvert at all, the threat of successful litigation appears minimal. However, if as in this case properties are threatened that would not be threatened with a natural watercourse (no culverts or freeway), the agency should assign upgrading a high priority.

8 Fire–Flood Sequence: A Deadly Combination

THE EVENT

The Southern California winter rainfall seasons of 1975–1977 were so devoid of precipitation, Los Angeles County officials reluctantly approved a cloud seeding program in an attempt to turn the spigot on.

In the San Gabriel Mountains' Angeles National Forest, only 20 mi (32 km) north of downtown Los Angeles, the drought had resulted in an extremely high fire hazard. The term national forest ordinarily conjures up a mental picture of tall pine trees and cool rushing streams—a false image for this particular locale. The only existing trees were those few that could survive on the most meager nourishment. Further, there was not a hint of moisture in the canyon bottoms. The thick chaparral and brush that clung to the rugged precipitous mountain slopes were tinder dry and highly combustible.

Into this charged environment, on July 24, 1977, wandered an antique gun aficionado firing a black-powder muzzle-loading rifle. His activity was perfectly legal because the area had been designated an open firing range by the U.S. Forest Service. However, on that day he would literally trigger a chain of events that would terminate 6.5 mo later in a terrifying and tragic calamity. According to a fire warden's report, the rifleman used facial tissue for wadding instead of nonflammable cloth. The tissue apparently smoldered and burned, quickly igniting a wildfire that totally consumed 3860 acres (1560 ha) in two watersheds. It took 3 days and $750,000 to bring the fire under control. By that time it had destroyed numerous facilities, including power lines and a highway patrol microwave repeater station. However, these losses proved insignificant in contrast to the loss of vegetative cover.

Figure 8–1 illustrates the boundaries of the fire and the two watersheds, Middle Fork and Mill Creek. The tiny Mill Creek resort community of Hidden Springs contained most of the residents and structures within the watersheds. In 1977 Hidden Springs consisted basically of four camp facilities (Fig. 8–1, inset).

It was obvious to both the Forest Service and the Los Angeles County Flood Control District that the denuded watersheds could pose a grave threat to inhabitants and facilities in the event of a high-intensity storm during the

186

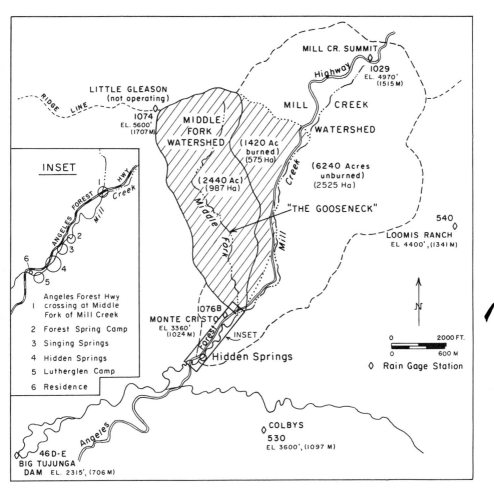

FIGURE 8–1
Map showing Middle Fork and Mill Creek Watershed.

next several rainy seasons. Consequently several steps were taken to help protect these camps as well as several private residences:

- The Forest Service reseeded the burned areas, knowing there was probably insufficient germination and growth time to help much in the next rainy season. The reseeding work was undertaken not only to lessen erosion of the burned watersheds over the next several years but to decrease siltation of the Big Tujunga flood control reservoir located about 5 mi (8 km) downstream.
- The flood control district conducted an investigation of the potentially

Typical condition of watershed contributing to Hidden Springs. Note denuded terrain and extreme steepness of slopes.

vulnerable facilities and structures, suggested to the owners that certain improvements be made, and, in general, warned the owners of the flooding hazard. (Mill Creek, at Hidden Springs, is on private property.)
- The local sheriff's office also was made aware of the possible need for hasty evacuation of the facilities in case of heavy rainfall.

Another action was taken by the Los Angeles County Road Department. It was to augment the 9 ft (2.7 m) diameter corrugated metal pipe culvert under Angeles Forest Highway at the Middle Fork crossing with an 8-ft (2.4-m) culvert and to excavate a modest debris settlement area upstream of the road. The purpose of this work was to help protect the roadway itself.

The 1977–1978 rainfall season wasted little time making up for the two previous dry seasons. Rainfall from mid-December to February 8 was double the normal. At Hidden Springs the total for the period was about 20 in. (51 cm) spread over many small and moderate nonintensive rainfall periods, and hence there was little flooding. However, the soil mantle had gradually become saturated, leaving little opportunity for further infiltration (10 in. [25 cm] of antecedent rainfall is generally all that is necessary to saturate the soil). The environmental setting was now complete for the impending disaster.

On Wednesday, February 8, after a day of respite, heavy rain began to fall over the entire Los Angeles basin. By the evening of the next day, another 2–3 in. (5–8 cm) had fallen in Hidden Springs, but still no appreciable flood damage had occurred. Then, between midnight and 1:00AM on February 10, the steady rain suddenly became a deluge. A high-intensity storm cell, starting offshore in the Pacific, swept northerly across the county in a path shown in Figure 8–2. The storm caused severe flooding throughout its path, but the effect on Hidden Springs was devastating.

The water in Mill Creek had been steadily rising for 2 days and, by midnight, had reached a point where the residents were becoming concerned. Ironically, at about midnight, as the storm began to peak, a fire broke out in the wash house at the Hidden Springs camp. Because the Forest Service had

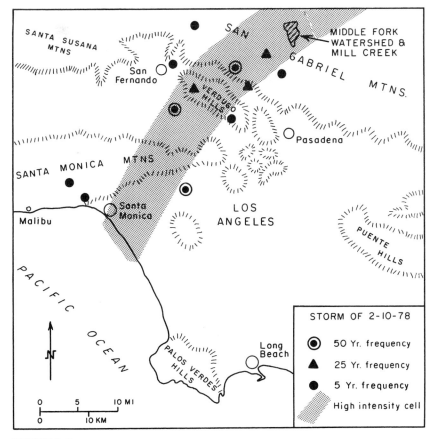

FIGURE 8–2
Storm path of February 10, 1978 (from Los Angeles County Flood Control District).

responsibility for fire suppression in the area, one of its employees, a supervisory tank operator, was awakened by telephone. He quickly aroused two others, his son (a firefighter) and a tanker operator. They jumped into a tanker truck and a patrol vehicle and headed for Hidden Springs. At about 1:50 AM, with the help of the rain, the fire was brought under control.

The firefighters were still mopping up at about 2:30 AM when the water became about 1 ft (30 cm) deep in the wash house. Minutes later disaster struck. The only warning began as a dull roar. Within seconds the noise was deafening. The fire fighters and bystanders looked up to see a wall of water about 15–18 ft (4.5–5.5 m) high bearing down on them. Laden with mud, rocks, logs, and other debris, the wave demolished and carried away nearly everything in its path. With virtually no time to react, everyone frantically reached for something to cling to. Some were fortunate. Most were not.

The supervisor, who was standing in the doorway of the wash house when the wave struck, was never seen again, nor was his body ever recovered. When his son saw the wall of water about to crash over him, he dived through a window into the laundry room. There he hung precariously to a window frame as the fire-gutted structure was swirled away. Somehow the building lodged itself against a large cottonwood tree and managed to withstand the

Path of destruction of the Hidden Springs floods and debris flow. The crest height of water in the channel area was estimated to be over 15 ft (4.5 m). The bulking factor was estimated at 400%.

surge. He and another survivor were trapped in what remained of the structure and were later rescued. Both sustained only minor injuries. The duration of the surge was only about 20 sec. It could not have been much longer because neither survivor was under water long enough to drown.

Meanwhile, the pump operator was first swept under the tanker, then beneath an automobile, where he lay submerged in water and mud. He was freed when the force of the surge lifted the car, but not before the wheels had rolled over his legs. Miraculously he was not badly injured and was able to scramble to high ground.

A deputy sheriff, who had been helping fight the fire, also was a fortunate survivor. Plummeted downstream, clinging to a door of the patrol vehicle, he managed to grab a tree branch and pull himself out of the torrent.

Within about 30 min after the wave surged through Hidden Springs, water in the creek had subsided to its prewave level. It took several days to determine that 13 people in the Hidden Springs area had lost their lives. Only one body was ever recovered. The tanker was never found, although some of its equipment, along with the shattered remains of the patrol vehicle, were discovered about 0.5 mi (0.8 km) downstream from the Hidden Springs camp.

The other three camps sustained severe structural and landscape damage, but no lives were lost.

THE INVESTIGATION

In February 1981, 3 yr after the disaster, the authors were contacted by the U.S. Attorney's Office in Los Angeles to determine our interest in joining a team of forensic experts.

We learned that the United States government was being sued by the relatives of the disaster decedents and the shareholders of the Hidden Springs camp. The plaintiffs' specific allegations of negligence against the U.S. Forest Service included:

1. Failure to analyze adequately the flood potential after the fire
2. Failure to rehabilitate the burn areas and control the runoff adequately
3. Failure to create an effective flood warning system.
4. Failure to warn the residents of Hidden Springs immediately prior to the flood
5. Failure to supervise and inspect adequately certain grading and culvert work done by the Los Angeles County Road Department under its federal use permit
6. Failure to consider the effects of cloud seeding on potential flooding before granting a permit to the county to do so

We also learned that the other defendant was Los Angeles County. Allegations against the county were both broader in scope and contained more specifics than those against the Forest Service.

We informed the assistant U.S. attorney handling the case that, if employed, our engineering and geologic investigations would cover all aspects of cause, effect, and prevention within our areas of expertise and would not be confined to addressing specific allegations. Further, items 1, 3, 4, and 5 were essentially nontechnical in nature, and item 6 was not within our technical province. She agreed that our investigation should encompass the broad view of what caused the disaster and if or how it could have been avoided.

In September 1981, as we were proceeding with our investigation, the U.S. attorney was successful in a motion for summary judgment, thereby eliminating the Forest Service as a defendant. Shortly thereafter we were employed by the Los Angeles County counsel to complete our investigation on behalf of the county. We discovered that, in addition to the same allegations directed at the Forest Service, the suit against the county also involved accusations of failure to take actions to control the flooding and charges related to road construction and damming.

We began our investigation with a general field reconnaissance followed by a review of all the existing discovery data, including maps, depositions, reports, and letters. Based on this information, plus what we had seen in the field as well as some very preliminary hydrologic and hydraulic calculations, we had already arrived at one firm conclusion: *The magnitude of this flood could not be explained on the basis of normal hydrologic criteria, including erosion bulking factors.* It was apparent that we must look for some highly abnormal or unusual conditions.

Topography

It is difficult to conceive of a forensic investigation of an environmentally related disaster that does not involve the use of topographic maps and surveys. In many instances such maps are crucial in the forensic process, providing the basis for hydrologic, hydraulic, and geologic analyses. Since it is often necessary to conduct these studies for previous time periods, obtaining accurate older maps and surveys becomes important. Sources in the United States for older maps include:

- USGS
- National Archives in Washington, D.C. (it is possible to give this organization geographic coordinates and receive a composite print-out of all maps within such coordinates)
- State archives
- State geological surveys
- University libraries

Sources for more recent topographic maps include local government offices, private aerial photogrammetry companies, and other civil engineers and surveyors. In the absence of specific current topographic mapping, it is often necessary to employ a surveyor and/or a photogrammetry firm.

The only topographic maps we could find covering the entire Mill Creek and Middle Fork watersheds were USGS quadrangle sheets. Since the county was able to furnish larger-scale topographic maps of the areas adjacent to Angeles Forest Road, initially we did not feel it necessary to do any field survey work. From the quadrangle sheets we planimetered the watershed areas draining to Hidden Springs:

Middle Fork (all burned)	2,440 acres	(976 ha)
Burned portion of Mill Creek	1,420 acres	(568 ha)
Unburned portion of Mill Creek	6,240 acres	(2496 ha)
Total	10,100 acres	(4040 ha)

The maps confirmed several things we had observed during the site inspection and a helicopter fly-over:

- The watershed slopes were extremely steep, particularly in the Middle Fork, where about 90% of the total 2440 acres (976 ha) had slopes in excess of 45 degrees.
- Despite an intervening 5 yr of higher than normal rainfall, the vegetation on the slopes of the Middle Fork was sparse compared with adjacent watersheds. (These first two observations led us to believe, early on in our investigation, that actual erosion had been much greater than theoretic.)
- The rock was an intensely shattered coarse-grained granite, obviously highly susceptible to erosion. Fracturing was related to a fault zone. The soil mantle was thin, consisting of extremely erodible coarse-grained gravel.
- There existed an extremely narrow segment of Middle Fork Canyon, located about 1 mi (1.6 km) upstream from Angeles Forest Road, aptly named "the Gooseneck".

In the narrowest portion of the Gooseneck there was evidence of high-water marks about 20–25 ft (6–7.5 m) above the canyon bottom. We immediately pondered the possibility that the Gooseneck might have become temporarily dammed with large rocks during the storm. As the water level built up behind the dam, the hydrostatic pressure might have dislodged the rocks, instantaneously disgorging a huge volume of water, mud, and debris. It was a theory we would investigate.

Photographs

The old adage about a picture being worth a thousand words is nowhere more applicable than in forensic work. Photographs not only are invaluable tools for investigation but serve as exceedingly effective court exhibits. The use of videotape for both purposes is growing so rapidly in popularity that few forensic civil engineers and geologists do not have a video camera. For example, videotaping serves as an excellent means of recording geologic features exposed in a boring or trench as well as recording rock and soil taken from continuous cores.

Aerial photographs are useful to both the civil engineer and the geologist. For the engineer they serve to supplement the topographic maps in the study of watershed characteristics that affect hydrologic and hydraulic calculations. For the geologist they serve a multitude of purposes, including the depiction of

- Terrain characteristics
- Faults
- Landslide evidence
- Sedimentation boundaries
- Erosion features
- Visual reference to geomorphic setting
- Changes that have occurred over time (using aerials of the same area taken in different years)

There are excellent sources of aerial photography available. One of the best is the U.S. Department of Agriculture's Soil Stabilization and Conservation Service. This service has compiled photography secured for various federal agencies through the National High Altitude Photography Program, begun in 1978. Coverage is comprehensive throughout the country and quality is excellent. Satellite imagery also is available through the U.S. Department of Agriculture. Both high-altitude aircraft and satellite photography are available through the EROS data center of the USGS. Recently the authors were involved in the forensic investigation of a landslide in which it was important to determine whether an irrigation reservoir located uphill from the landslide had been filled with water during a certain time frame. Since there was conflicting testimony among the principals and eyewitnesses, satellite imagery would have been sufficiently accurate to resolve the question. Unfortunately in this particular case we were denied access because of security reasons, the site being only about 3 mi (5 km) from then President Reagan's Santa Barbara ranch.

It is possible to order a computer geographic search of aircraft coverage through the nearest National Cartographic Information Center Office of the USGS merely by submitting the latitudinal and longitudinal coordinates of the area under investigation. This is a tremendously useful service in locating aerial photography.

When aerial photography is not available from government sources or private aerial photogrammetry companies, it may be necessary to have it flown for the specific forensic purpose. Care should be exercised to specify the desired scope and scale.

In this Hidden Springs investigation we were able to secure a wealth of useful photographs, both aerial and ground, from a variety of sources. Because of the suddenness and ferocity of the flash flood, we could find no photographs taken during the event.

Rainfall and Hydrology

Nearly all environmentally related forensic assignments require the extensive use of precipitation records, including statistical data. In addition to aiding hydrologic studies, which could not be made without them, precipitation records are an important element in the analysis of all earth movement, including landslides, mud flows, subsidence, and other ground/slope failures.

Sources of local rainfall and snowfall records vary considerably throughout the United States. Some local jurisdictions have established comprehensive networks of rainfall gages, including many that record automatically at short time intervals. Where insufficient precipitation data are available at the local level, state agencies responsible for public works, highways, geology, and water planning should be contacted. Finally, other federal sources include the U.S. Weather Bureau, the Soil Stabilization and Conservation Service, and the USGS.

During heavy precipitation the automatic gages are extremely valuable for hydrologic calculations because their records permit direct correlation to time of concentration. Those that are not automatic are usually read by agency personnel on a daily basis during a storm period.

A great deal of statistical information is also available at some local government agencies. This includes isohyetal maps, intensity/duration curves, return frequency charts, and historical rainfall records. Some agencies have mapped their entire jurisdictional area showing rainfall intensity zones, isohyetal lines, soil types, debris potential, and infiltration rates.

It is desirable to obtain precipitation information from as many sources as possible and to compare the results carefully. Meteorologic data for any given storm would indicate that rainfall intensity along the storm cell paths is not uniform. Local squalls, for instance, can produce intensities of much different magnitude at locations only 1 mi (1.7 km) apart. Thus it is important to examine all available rainfall data before estimating intensities for the watershed being investigated.

Stream gages are often under the same jurisdictions as rainfall gages. Before embarking on a stream discharge calculation it should be determined whether there was a stream gage working at the time of the disaster. This seems elementary, but hydrology experts have sometimes merely assumed that none existed.

Forensic projects adjacent to the ocean often require tidal data. Some local jurisdictions record such data but generally it is necessary to contact the U.S.

Weather Bureau, the U.S. Army Corps of Engineers, the U.S. Navy, or state water resources agencies.

The objective of our hydrologic analysis in this case was to roughly estimate the theoretic peak flow, based on rainfall records, and to compare this discharge with the actual volume correspondng to the height of water witnessed at Hidden Springs camp (15–18 ft: 4.5–5.5 m) and to other high-water marks in the creek.

A summary of pertinent information derived from records of nearby rainfall gages is shown in Table VII. After reviewing the records and the storm cell path data we decided to use Station 46-D-E, which was most comparable to rainfall intensities within the Middle Fork watershed. For the Mill Creek watershed we used a combination of Stations 54C, 1029C, and 53D. In order to estimate roughly the theoretic peak flows at various points, we employed a simple Rational formula approach. The times of concentration were estimated to be:

1. Middle Fork watershed to road crossing 45 min
2. Mill Creek watershed to road crossing 50 min
3. Road crossing to Hidden Springs 8 min

The times of concentration for each watershed were sufficiently close to assume that they were identical (50 min to road crossing); thus their peak flows could be added. It was already obvious that the return frequency of the storm, in terms of the time of concentration, approached 25 yr. The coefficients of runoff for each watershed, burned and unburned, were estimated on the basis of criteria formulated by the Los Angeles County Flood Control District.

The following table summarizes the theoretic peak flows:

Peak flow from Middle Fork only at road crossing		Combined peak flow from both watersheds at Hidden Springs	
With unburned watersheds	With actual burned watersheds	With unburned watersheds	With actual burned watersheds
1600 ft³/sec (48 m³/sec)	6100 ft³/sec (184 m³/sec)	3000 ft³/sec (90 m³/sec)	8800 ft³/sec (264 m³/sec)

The theoretic difference between peak flows for the same storm, burned and unburned, is strikingly exemplified by the Middle Fork Qs at the road crossing:

$$\frac{Q \text{ with burn}}{Q \text{ with no burn}} = 3.2$$

TABLE VII
Rain Gage Summary

County gage station	Location	Rainfall total storm		Rainfall maximum 24 hr		Approximate return frequency (yr)	Rainfall maximum 1 hr		Approximate return frequency (yr)
		(in.)	(cm)	(in.)	(cm)		(in.)	(cm)	
46D-E	Big Tujunga Dam elev. 2315 ft (695 m)	11.6	29	9.9	25	>5	1.6	4.1	25
47D	Clear Creek Schools elev. 3100 ft (930 m)	14.5	37	12.3	31	>25	1.4	3.6	>5
53D	Colbys elev. 3620 ft (1086 m)	15.5	39	11.5	29	30	1.1	2.8	>5
57B-E	Camp Hi Hill elev. 4250 ft (1275 m)	12.3	31	11.0	28	25	0.6	1.5	1
54 C	Loomis Ranch elev. 4300 ft (1290 m)	9.5	24	6.2	16	5	0.6	1.5	1
1029C	Mill Creek Summit[a] elev. 4970 ft (1491 m)	8.5	22	5.1	13	<5	0.4	1.06	<5
1029B	Tujunga Mill Creek[b] elev. 4600 ft (1380 m)	7.7	20	5.1	13	<5		[b]	
1074	Little Gleason elev. 6100 ft (1830 m)			Gage failed—No data					
1076B	Monte Cristo Ranger Station[b] elev. 3360 ft (1008 m)	11.54	29	8.9	23	10		[b]	

[a] U.S. Weather Bureau station.
[b] Standard gage; no hourly information available.

Using the Manning formula with slopes and areas derived from county topographic maps, we calculated the theoretic maximum water-surface height generated by a Q of 8800 ft^3/sec (264 m^3/sec) at Hidden Springs to be about 8 ft (2.4 m). This figure was considerably less than the wave height described with unanimity by the eyewitnesses (15–18 ft; 4.5–5.5 m), and confirmed our original theory that the devastating surge could not be explained on the basis of conventional hydrologic criteria.

We investigated what effect a temporary damming of the Gooseneck might have had on the peak flow. Using the high-water marks that were plainly visible, we calculated the maximum volume of water and debris that could have been stored and then suddenly released. Because of the steepness and narrowness of the canyon at this point, the storage volume, even if suddenly released, would not have been enough to have a meaningful effect on the surge.

A review of the literature on roll waves (or bores) revealed that the conditions were perfect for the formation of such a wave or waves. These conditions included an extremely rapid increase in debris flow plus what must have been violent turbulence at the confluence of Middle Fork and Mill Creek. The imposition of a roll wave on top of the normal theoretic peak flow would certainly account for a portion of the difference in water level. However, the

"Gooseneck" curve where possibility of temporary blockage was investigated. Highway can be seen in upper left corner of photograph.

most important factor in explaining the 15–18 ft (4.5–5.5 m) surge was the highly erodible nature of the particular soil and bedrock contained in these watersheds.

Geology and Geotechnical Engineering

Our research of pertinent technical papers and other empirical data revealed conclusive evidence that erosion potential increases dramatically after a wildfire as the result of removal of the vegetation and its litter. The grass, brush, and trees not only provide a canopy to protect the soil from rainfall impact but also absorb a portion of the rainfall for evapotranspiration. In addition, litter acts as a series of microdebris basins on a slope, retarding the effects of sheetflow.

High-temperature brush fires of about 400°F (204°C) change the infiltration capacity of the soil near the surface and the permeability of the subsurface soils. As the nature of the soil profile is altered, fine-grained materials tend to move downward through the soil profile causing a shift in particle size distribution, with the near-surface soils becoming coarser grained and more porous while the lower soils densify as the result of the increasing percentage of fines. This change in particle size distribution reduces the permeability of the lower, denser soils.

In addition to the effect of particle size redistribution, a water-repellent layer develops at 2–3 in. (5.1–7.6 cm) below the surface owing to the concentration of waxlike complexes of long-chain hydrocarbons derived from the burning of vegetation. Origin of this repellent zone is associated with the burning of litter and overlying vegetation whereby the vaporized hydrocarbons move downward and condense where the soil is cooler. The ensuing coating of the soil particles produces a subsurface water-repellent layer. The effect of this water-repellent layer is to reduce the storage capacity of the soil mantle by at least 20 times. Thus, during a high-intensity storm the top 2–3 in. (5.1–7.6 cm) of surface soil is rapidly filled, causing runoff to begin sooner and runoff rates to be much higher. Studies indicate that 1 in. (2.5 cm) of rainfall on a burned watershed can be as damaging as 5 or 6 in. (13 or 15 cm) on an unburned watershed.

In the case of the Middle Fork, the geologic conditions accentuated the effect of the increased runoff on erosion and sediment production. The bedrock is a highly sheared cataclastic anorthosite that has been intensely fractured by past faulting and converted by fracturing and decomposition to a coarse, granular, decomposed granitic soil. Most of the gravel-sized fragments are individually nearly rectangular-shaped feldspar crystals. Soil thicknesses range from 6 to 18 in. (15 to 46 cm) on hillside slopes. These soils are very erodible under unburned conditions and are subject to dramatic increase in erodibility following a fire. Research has shown that erosion may be increased 35 to 48 times resulting from high-intensity runoff during the first year after a fire.

Immediately following a fire, gravity becomes a dominant factor affecting the surficial soils; dry ravel occurs, moving loose materials downslope onto the lower portions of the slopes and into the canyon bottoms. Later, during the ensuing winter rain, runoff begins moving the loose materials that have collected in the canyon bottoms. Gradually the loose material on the lower portions of the slopes begins to wash downslope into the canyon bottoms and becomes involved in the sediment transport process. Because of the lack of vegetation and litter to retain the loose soils, the erosion rate is greatly exaggerated and the bulking factor increases dramatically. The greatest increase in erosion and sediment transport occurs after the soils have been saturated from many days of antecedent rainfall and then impacted by torrential rain associated with high-peak-intensity rainfall. All these conditions existed in the Mill Creek/Middle Fork watersheds.

Using criteria developed by the Los Angeles County Flood Control District, we found the theoretic bulking factors for the Middle Fork watershed to be 2.0. In other words, for a capital (50-yr) storm on this freshly burned watershed, the bulked Q would be double the clear water flow. However, the hydraulic calculations based on actual high-water marks indicated a Q of three to four times the clear water flow. This result reinforced our theory that the specific, unusually high erodibility reflected in the geologic characteristics of this watershed was simply not fully recognized in the more generalized approach used by the county in establishing its criteria.

It appeared to us that the extreme erodibility of the burned watersheds and the development of a roll wave were the most plausible explanations for the surge.

The sudden increase in runoff was accompanied by a corresponding increase in velocity. Erosion and transport capability are, in turn, increased in geometric proportions to velocity increase. Supportive evidence of this tremendous velocity was the existence of 2–5-ft (0.6–1.5-m) diameter boulders that had been transported. The veolcity necessary to move these boulders is in the range of 20–30 ft/sec (6–9 m/sec).

Angeles Forest Highway

The plaintiffs alleged the highway acted as a dam during the flood and that, when it was breached, the flow to Hidden Springs was suddenly and dramatically increased. A review of numerous photographs taken by the Forest Service the day after the flood revealed the road fill had not been breached. This fact was substantiated in the depositions of eyewitness county road department personnel.

Obviously, at the time of the surge, the two culverts lacked the capacity to handle the peak flow of 6100 ft³/sec (183m³/sec) versus 2000 ft³/sec

Debris covering the Angeles Forest Highway after Hidden Springs flood–debris flow. Note the size of the boulders visible in the lower right. The level area where vehicle is parked and upstream to the left represent the filling of a canyon with sand and boulder-sized debris. The roadway was not breached, but the debris-laden flow across the roadway was estimated at 380,000 yd³ (304,000 m³).

(60 m³/sec), which was the combined capacity of the culverts. The excess Q built up and overtopped the road. High-water marks indicated a maximum water-surface height above the road of about 6 ft (1.8 m). Using this depth and treating the road section as a weir, we computed the maximum Q over the road at about 6000 ft³/sec (180 m³/sec). This confirmed our other calculations.

We therefore concluded that the existence of the county road did not increase the peak flow at Hidden Springs. In fact, the small volume of road fill that was eroded was offset by the volume of storage upstream of the road created by the road department when it installed the second culvert.

Mitigation Measures

We carefully explored what physical work might have been done in the 6-mo period between August 1977 and February 1978 to mitigate the effects of the flash flood on Hidden Springs.

Since the flood was essentially a debris flow, it would have been necessary to capture a large portion of the debris. Basic mitigation would have consisted of excavation of a large debris basin, construction of an earth-fill dam, utilizing the excavated material from the basin, and installation of a perforated standpipe and storm drain to filter the clear water and discharge it downstream. We assumed the design volume of the basin to be 380,000 yd^3 (304,000 m^3), which was a figure predicated on flood control district standards and criteria but which we thought was low.

Two possible sites for the hypothetic facility were examined: one immediately upstream from Angeles Forest Highway (Fig. 8–3), and the other at the Gooseneck. In order to retain even 380,000 yd^3 (304,000 m^3), the Gooseneck dam would have to have been about 160 ft (48 m) high, putting it in a "major dam" category. We rejected the Gooseneck as a potential site immediately. Our focus turned to the road site with the following results (see Fig. 8–3):

1. Approximate height of dam = 50 ft   (15 m)
2. Aproximate length of dam = 450 ft   (135 m)
3. Approximate area of debris basin = 15 acre   (6 ha)
4. Cost estimate (1978) = $2.5–3 million
5. Estimated minimum time to complete:

   Geologic and geotechnical investigation = 3 mo

   Design = 2 mo

   Approvals = 3 mo

   Construction = 6 mo

   Total = 14 mo

The approvals would have been particularly time-consuming because the California State Division of Safety of Dams must review and approve all plans and specifications for dam heights exceeding 25 ft (7.5 m) and dams that impound more than 15 acre ft (1.8 ha m). Even if one were to assume that, by some miracle of fast-tracking, the dam could have been designed and built in 6 mo, the cost could not have been justified in terms of the value of the land and improvements it would protect. Not even included in the cost estimate is the postflood expense of removing, transporting, and depositing the debris to a suitable dump site, which are formidable and expensive tasks.

We concluded that there had been no feasible method of mitigating the disaster through physical means.

The results of our investigation, which were of primary importance to this case, could be summarized as follows:

1. Owing to the fire–flood sequence, certain characteristics of the soil topography, and rock type, the erosion rate was dramatically higher than normal.
2. Because of abnormally high antecedent rainfall the denuded soil mantle

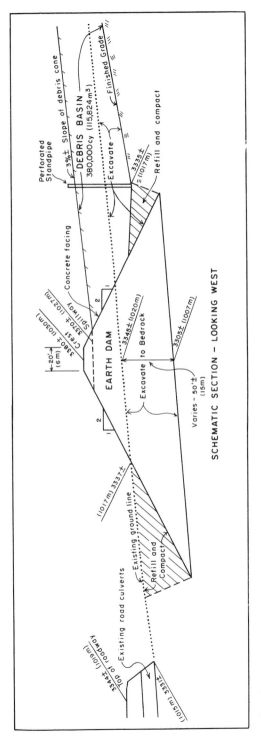

FIGURE 8-3

Hypothetic debris basin for middle fork at Angeles Forest Highway.

was so saturated at the time of peak intensity that the runoff factor was nearly 1.0.

3. The sudden surge of sediment-laden flow was added directly to the nearly full bank flow. This abrupt increase in water surface created a gigantic roll wave.

4. The roll wave superimposed on the peak bulked flow created the 15–18-ft (4.5–5.5-m) surge that devastated the Hidden Springs community.

5. The rainfall, erosion, and resulting flash flood constituted an event that was caused and controlled by the natural setting (except, of course, for the ignition of the fire).

6. There was no evidence of man-made damming effects, with ensuing failure of the dam, that could have contributed to the magnitude of the flood.

7. There were no feasible methods available to mitigate the effects of the flash flood.

THE LITIGATION

Two separate legal actions were filed against Los Angeles County: one by the owners of the Hidden Springs camp and relatives of the decendents, and the other by the owners of The First Evangelical Lutheran Church of Glendale (Lutherglen). Both lawsuits claimed liability damages as well as inverse condemnation. In this case the inverse condemnation claim by the plaintiffs' attorneys arose from the county's alleged failure to provide adequate flood protection.

The Lutherglen case was the first to come to trial. Depositions of the expert witnesses from both sides revealed that there was little difference of scientific opinion as to the cause and general magnitude of the surge. It became apparent early on that the basic technical question in dispute was going to be whether the county could have taken any reasonable action in the 6.5-mo period between the fire and the flood to mitigate the effects of the flood.

Lutherglen's attorneys appeared to build their case on basic allegations that the county controlled Mill Creek and therefore should have:

1. Known that the hazard existed
2. Known how to eliminate or mitigate the hazard
3. Taken action

There was no dispute about whether the county recognized the hazard. Personnel from the flood control district were dispatched after the fire to warn residents of the potential for flooding during the coming rainy season. No one, of course, could foresee the magnitude of the storm that occurred.

Based on its own bulking criteria, the county recognized the potential for at least 380,000 yd³ (304,000 m³) of silt, rock, and debris in the event of such a storm. As previously described in our investigation, there appeared to be no way the county could have mitigated the disaster given the short time to design, process approvals, and build a dam and debris basin. Even if it could have been done, it would not have been cost-effective and would have created an even greater potential hazard in the event of dam failure.

Oddly, one of us was subpoenaed by the plaintiffs' attorneys even though they did not call their own engineering experts. The only explanation we could derive for this legal maneuver is that they hoped we would help their case more in direct examination than in later cross-examination.

The strategy proved futile because the trial never reached the point of defendants' testimony. When the plaintiffs finished their testimony, the county counsel submitted a brief to the court requesting a nonsuit. The judge granted it.

With the trial concluding so quickly, there was a unique opportunity for the participants to talk with the jury members. We were impressed with their understanding of the issues and interested in their comments and observations regarding our testimony. There is no better gage of quality of performance and degree of credibility than from the jury.

Much can be learned and stored away for future reference when the trial has been completed. For instance, the court retains the exhibits only long enough to determine if there is going to be an appeal. At that point the experts, through their attorney clients, can retrieve their own exhibits. The exhibits can prove extremely helpful in future cases. Environmentally related lawsuits, like history, have a tendency to repeat themselves.

The plaintiff appealed the decision, but the appellate court denied the appeal. The other lawsuit by Hidden Springs et al. never came to trial. We and the plaintiffs' experts were deposed, but apparently no new evidence or theories surfaced during these depositions that would give the litigants cause to think they could do any better than Lutherglen.

An interesting legal by-product emerged from this case that may be important to land-planning agencies and public administrators throughout the country. Having lost its case, the owners of the Lutherglen camp applied to the county for a permit to rebuild its damaged facilities. To thwart the possiblity of another disaster, the county had adopted interim floodplain regulations for the Mill Creek floodway that precluded any new structures. The permit application was therefore denied. (The regulations did not affect the use of existing facilities that had not been damaged.)

Lutherglen sued the county in Superior Court, alleging that the property was, in effect, being taken without compensation. The defendant county won the case. Lutherglen appealed unsuccessfully, first to the California Court of Appeals and then to the California State Supreme Court. Next came an ultimate appeal to the U.S. Supreme Court. In a 5–4 decision the court ruled for Lutherglen. The court majority was quick to point out that its decision should not be interpreted as questioning the broad constitutionality of floodplain management practices or ordinances, but, rather referred only to the narrow circumstances of this particular case. Although the decision initially caused great consternation in the floodplain management community, a thoughtful and knowledgeable legal appraisal of its consequences has generally concluded that land-use regulations, per se, do not constitute a taking. One wonders what the result would have been had the county, instead of enacting an all-prohibitive zoning ordinance, merely required the owner to submit for approval reports and calculations by licensed professionals as to the ability of the planned facilities and its occupants to be safe in the event of a fire–100-yr-flood event. Undoubtedly such construction would not have been cost-effective. But if the applicants still wanted to proceed, the county would not have ended up with substandard and/or improperly sited facilities.

Despite its legal victory Lutherglen has not rebuilt as of this writing.

THE POST MORTEM

Soaring population growth in the fire–flood-prone western states increases pressure to develop mountain and foothill areas more intensely. Much development has already taken place; some incorporates adequate protective controls, but most possesses some degree of risk exposure. Especially vulnerable are structures in relatively remote areas, where there has been little or no thought given to such environmental problems and where, in many cases, no building permit of any kind was obtained, maybe not even required, at the time of construction.

As a consequence, serious questions arise for engineers and geologists concerned with development in regions subject to fire–flood phenomena. Among them are:

1. What can be done to protect structures (and their inhabitants) already improperly sited and built relative to potential debris flow and flash flooding (as in this case)?
2. How can engineers and geologists work with land-use planners and administrators to assure that future structures and facilities will be properly sited and protected?
3. What criteria should be used to evaluate most accurately debris production potential?
4. What are the legal implications professionals should consider in planning for fire–flood protection?

The obvious structural solutions to the problem of protecting existing improvements are to build dams and debris basins to trap the mud and debris and open channels or storm drains to convey the clear water to suitable outlet facilities. However, as exemplified by this case, the cost of such structures may be far in excess of the value of the improvements to be protected.

Certainly it is incumbent upon appropriate public entities to examine all existing development within their jurisdictions subject to fire–flood damage and to make recommendations for migitation of the hazard. Recommendations would run the gamut from:

- Condemnation/purchase and demolition of the most vulnerable properties in sparsely developed areas and the establishment of a flood control easement or right-of-way

to

- Establishment of an assessment improvement district composed of benefitting property owners in order to finance the construction of debris basins and storm drains

to

- Construction of needed facilities with public funds in more heavily populated areas or where assessment districts are impractical

to

- Recommendations to individual property owners that certain added construction or reconstruction might lessen damage from storms of lesser magnitude

to

- Do nothing except establish some type of early warning system

It is obvious that, in most cases, one cannot wait until the fire occurs to start making plans for the flood. It is simply too late. Certainly there are specific actions to be set in motion immediately after the fire that include a reseeding program, formal warnings to residents and owners who could be affected, and the construction of whatever type of improvements seem appropriate. It goes without saying that the existing debris basin should be cleaned out and, if feasible, deepened and/or made larger to accommodate more debris. To condemn and demolish facilities, even those that are extremely vulnerable, appears to stretch the doctrine of eminent domain to its limit of credulity. How far should government go to protect people who could protect themselves by exercising a little caution and common sense? Despite the political and philosophical questions raised by such condemnation, it is a fact that in certain situations it may be the most cost-effective solution, in addition to preventing potential injuries and loss of life. Whether it is cost-effective depends, of course, on the cost of litigation in connection with the condemnations and the potential cost of being sued if nothing is done and a disaster occurs.

It does not appear prudent to depend too heavily on advance warning as a means of protection. First, high-intensity squalls can move so quickly and unpredictably that, even with a well-planned procedure, there is probably too little time for evacuation. Second, during the developing phase of the storm, access by warning vehicles to these areas may be difficult and time-consuming. Communication by telephone may be undependable at such times, too. But probably most important, there appears to be an inherent stubborn streak in the human psyche that prevents many people from heeding warnings. For example, one of the Hidden Springs residents was warned by flood control district personnel of the potential for disaster soon after the fire occurred. The resident replied, in effect, that he had not evacuated during the storm of January 1969 and that he would ride another one out of necessity. (Actually the 1969 storm produced a greater 1-hr duration intensity than the 1978 storm, but there had been no fire in the watershed.)

In planning future development, programs and procedures can be inaugurated to minimize property damage and personal injury and loss of life from fire–flood sequences:

- Prepare comprehensive topographic mapping, delineating watershed boundaries and existing developments.
- Perform a geologic investigation of the entire area to determine component soil/rock erodibility in the event of fire. This investigation is extremely important. Erodibility can vary to a great degree depending on rock type and condition, soil mantle, vegetation, topography, and other factors. Estimates should not be predicated on rules of thumb.
- Based on the results of the geologic study, estimate the maximum bulked flows in the canyon for a 100-yr storm occuring within the first year after a fire. (This may seem conservative, but we think it is appropriate.)
- Plot the boundaries of the 100-yr-bulked flood along all canyon bottoms and foothill alluvial fans.
- Calculate total debris production for each canyon.
- Analyze the potential locations, configuration, and estimated cost of dams and debris basins to trap these volumes.

Armed with this information, informal planning decisions can be made as to where such improvements are desirable and cost-effective in terms of future developments. A basis for prioritizing and financing these improvements has also been provided.

Conversely, the information would indicate where development should be severely limited if not prohibited. Within the 100-yr floodways, strict controls can be enacted to prevent development without complete and comprehensive mitigation plans by competent professionals.

The process just outlined is not easily accomplished. The studies themselves need to be funded promptly and adequately. The plan must be sold to the community at large and the means found to finance construction, typically either by developers or by assessment districts. Because of the inevitable pressure brought by developers, individual building permit applicants, and property rights advocates directly on local politicians, the process will never work perfectly. However, to avoid disasters like that of Hidden Springs in the future, a program of this or similar nature must be undertaken.

When an existing debris basin proves inadequate in a fireflood disaster, attorneys for plaintiff downstream property owners begin asking two basic questions:

1. Was the design of the facility consistent with current generally accepted criteria?
2. Had there been reasonable maintenance of the basin to assure near maximum capacity at the time of the flood?

If the agency in charge is always prepared in advance to answer these questions affirmatively, its liability exposure will be minimized.

LAND SUBSIDENCE
ON A LARGE SCALE:
DIRE CONSEQUENCES

THE EVENT

When a community sits on tidal flats flanked by two rivers having inadequate upstream flood control facilities, periodic flooding can be expected. If over a period of time the average elevation of the community is lowered from 2 ft (0.6 m) above sea level to 4 ft (1.2 m) below, without compensating flood protection, disaster is openly invited. Such was the fate in 1983 of Alviso, an historic Hispanic residential section of San Jose, California, situated at the southernmost extremity of San Francisco Bay.

Alviso was not always part of San Jose. Its independent origin dates back to 1777, when it was chosen as the site for a Spanish mission. In the early 1800s Alviso became a seaport, serving not only the mission but many of the farmers who had established fruit orchards throughout the Santa Clara Valley. Because of the port facilities, light industry was also attracted to Alviso. But it was mainly agriculture that continued to expand and diversify, providing employment to many Alviso residents.

Another industry known as salt harvesting was initiated in the shallow waters of the bay just north of Alviso. It proved to be such an ideal location that the industry still exists today. The harvesting technique is exceedingly simple. Enclosures in the form of earth levees are built to trap seawater and prevent new seawater and fresh-water drainage from intruding. After the sun and wind accomplish their evaporative work, the residual salt is scooped out and processed. A gate valve is then opened, allowing seawater to enter the enclosure, and once the valve is closed the cycle begins anew. What made this portion of the bay so attractive for salt harvesting was its isolation, shallow depth, and relative freedom from severe wave action.

Unfortunately the levees that prevent seawater intrusion also inhibit fresh-water surface flow from reaching the bay. This, as later described, was one of many factors contributing to Alviso's flooding susceptibility.

By 1852 Alviso had grown to the point where it incorporated. However, its role as a port was greatly diminished in 1864 when railroad service began.

The Santa Clara Valley remained primarily agricultural until after World War II, when growth burgeoned. San Jose, in particular, was transformed from a farm seat and food processing center to an electronics industry giant, hub of what was to become known later as Silicon Valley. By 1970 its population had surpassed that of San Francisco and, because of its seemingly unrestrained urban sprawl, was being unflatteringly characterized as the "Los Angeles of the north."

The rapid urban growth in the valley contributed in at least four significant ways to Alviso's vulnerability to flooding:

1. The increased demand for water was met by excessive groundwater extraction, resulting in valley-wide land subsidence and dire consequences for the Alviso area.
2. In order to maximize groundwater recharge and minimize the necessity of importing expensive water, certain policies and procedures were adapted relative to the operation of dams and reservoirs that severely penalized flood control.
3. The change from rural to urban upstream land uses resulted in markedly higher storm-runoff discharges, making existing channel and levee systems grossly inadequate.
4. In 1969 Alviso was chosen as the site for a hugh new wastewater treatment plant to serve San Jose and Santa Clara. As a result nearly 1200 acre (480 ha) were protected by levees, blocking natural flow patterns, removing surface storage from the distributing channels, and directing greater volumes of flow toward Alviso.

These, as well as other consequences of the valley population boom, are explained in greater detail in The Investigation section of this chapter.

Alviso has been subject to flooding since its inception. Upstream development only intensified and exacerbated the hazard. The mission itself was flooded only 7 yr after its construction had been completed. Rather then rebuild, it was wisely decided to build a new mission in nearby Santa Clara. In terms of agricultural use the early floods were actually a mixed blessing. These early losses more often than not were offset by enhancement of the soil fertility through deposition of rich silt. Further, the storms provided additional water for the storage reservoirs, which, at that time, were used primarily for irrigation.

From when the mission was destroyed in 1784 to 1982, records indicate, that there were at least 10 major flooding episodes in Alviso. Sometimes the source of flooding was Coyote Creek, other times the Guadalupe River; in at least three instances both watercourses spilled over. Although flooding levels in Alviso were often of serious consequence (as high as 5 ft [1.5 m]), they subsided rather quickly, enabling the residents to return home within a day or two to face the task of cleaning and repairing.

In 1969 Alviso was annexed to San Jose, ostensibly so that San Jose could exert complete political and land-use control over its new nearby wastewater treatment facility. For Alviso residents the opportunity to be within the jurisdiction of San Jose held the promise of better community services, particularly protection from flooding as reportedly promised by San Jose officials.

By 1980 the population of Alviso had stabilized at about 1700. Most of the homes were single family and, while many were old and flood-battered, they were by and large fairly well maintained.

Figure 9–1 is a schematic regional map, and Figure 9–2 focuses on the immediate Alviso area. Shown are important physical features affecting the ability of Alviso to fend off flooding. They include Coyote and Anderson Reservoirs, Coyote Creek, and Guadalupe River, including their levee systems, floodgates, salt harvesting ponds, and the wastewater treatment plant.

From 1969 to 1982 no flooding of consequence occurred in Alviso. The 1981–1982 rainfall season was sufficiently heavy to saturate the soils and fill the reservoirs but brought no flooding until March 31. On that day a storm of

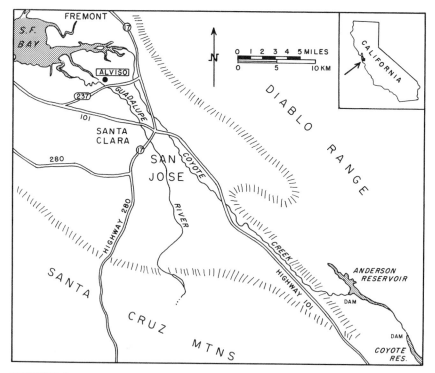

FIGURE 9–1
Regional map.

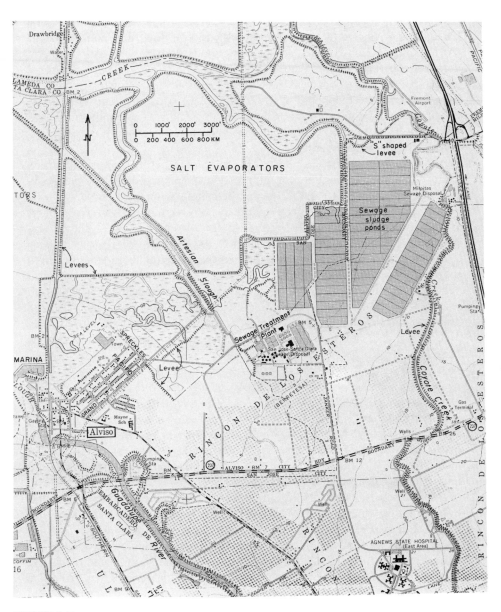

FIGURE 9–2
Map of Alviso and vicinity.

considerable magnitude caused six watercourses in the valley, including the Guadalupe River (but not Coyote Creek), to overflow their banks. As the storm continued on April 1, an apparent sudden release of water from Anderson Reservoir created a surgelike discharge in Coyote Creek. The surge caused five breaks to occur in the Coyote Creek levees, allowing floodwater to inundate Alviso to a depth of about 4 ft (1.2 m). Most residents voluntarily evacuated and did not return home for 4 days. Damage was estimated at $4.5 million and, in the aftermath, San Jose announced it would erect a ''ring'' levee around the community.

Alviso residents had less than a year to recover before misfortune struck again—this time with a flood of unprecedented magnitude. Chronologically the events leading up to and following the calamity ran this course:

1/29/83   As the result of near-record antecedent rainfall, Anderson Reservoir attained full capacity and began to spill over the dam. This spillover was to continue at varying rates for the next 67 days.

2/27/83   A high-intensity storm caused the spill rate at Anderson Dam to reach 16,000 ft^3/sec (480 m^3/sec). Coyote Creek overflowed its banks at several locations, but this resulted in only minimal flooding in the Alviso area.

2/28/83   The storm continued (about 3 in. [7.6 cm] of rainfall in Alviso and much greater in the upstream sections of the watershed) but was fairly steady and did not result in the peak discharges that cause bank overflows or levee breaches.

3/1/83   As the rainfall intensified, Alviso residents, with no ring levee in place as promised by the City of San Jose, hastily built one of their own by use of sandbags. Their efforts would prove to no avail. By noon Coyote Creek was overflowing its banks just south of the wastewater treatment plant. At 1:30 PM a temporary levee at Grand Boulevard and Spreckles Avenue was breached, resulting in 2 ft (0.6 m) of water in Grand Boulevard. At 2:30 PM high tide was about 9.1 ft (2.7 m). This condition inhibited storm flow to the bay, raising water levels upstream. At 5:15 PM the floodwaters breached one of the improvised levees at Grand Boulevard and Pacific Avenues and within a few minutes rushed into Alviso. Residents were quickly evacuated and before 2 hr had elapsed most of the community was under 4 ft (1.2 m) of water. It was the first mandatory evacuation in Alviso's history. Ironically, as late as 3:30 that afternoon a City of San Jose official had informed the media that he foresaw no problems from the rain-swollen Coyote Creek. Another interesting fact was that Coyote Creek had overtopped its banks at several points between Agnew State Hospital and the Fremont Airport, but the overtopping was of a sheetflow nature rather than due to levee or bank failure.

3/2/83   Rainfall in the Alviso area peaked that day at nearly 5 in. (12.7 cm). A major Coyote Creek levee was breached on the west bank near Agnew State Hospital, and water continued to pour into Alviso. By evening, flood depths in some of the lowest areas had reached a record 9 ft (2.7 m). At

Aerial photograph showing community of Alviso (left center), salt evaporators (left side and top), sewage ponds and plant (upper right), and Guadalupe River (from left center to lower right). Coyote Creek is off photo to the right (east). At peak of flood, water ranged from 3 to 5 ft (1 to 1.5m) deep. Floodwater entered Alviso from the south–southeast (lower right) at the same time the tide was rising in the bay. See Figure 9–2 for location of features.

5:30 PM a peak discharge of 5600 ft^3/sec (168 m^3/sec) was recorded by gage in the lower reaches of Coyote Creek. Statistically it represented about a 10-yr return frequency flow. A 100-yr flow (generally used for design purposes) is 14,500 ft^3/sec) (435 m^3/sec).

3/3/83 The City of San Jose had installed 14 pumps in Alviso with a combined output of 65 million gal (246 million liter) per day (100 ft^3/sec [3 m^3/sec]) to siphon water into the bay. However, floodwater was still flowing into Alviso at an average rate of about 350 million gal (1327 million liter) per day (540 ft^3/sec [16 m^3/sec]) despite the fact that rainfall in the basin had subsided during the day and had ceased entirely by evening. The explanation was that Anderson Reservoir was continuing to spill over its dam at an extremely high rate of flow.

3/4/83–3/6/83 Sometime during this period pump outflow became greater than flood inflow, and aided by a lack of significant rainfall, flood levels in Alviso began to subside gradually. Nearby areas that had also been flooded were nearly drained by March 6. It was obvious that the makeshift ring levee around Alviso had not only failed to prevent intrusion of floodwaters but had severely hampered their removal by natural drainage.

3/9/83 About 3 in. (7.6 cm) of rain fell over the entire basin, causing Alviso a minor setback. By the end of this day water levels stood at 2–3 ft (0.6 to 0.9 m) over the easterly and central sections of Alviso and 4–5 ft (1.2 to 1.5 m) in the westerly portion.

3/15/83 Water had subsided to the point where the Alviso residents were allowed back to their homes to start repair and cleanup work. In another touch of irony, this was the day the City of San Jose built a more substantial ring levee of earth and rock. the project was completed in 48 hr at a cost of about $50,000. Total cost of the physical flood damage to 375 homes and 40 businesses in the Alviso area was estimated to be in excess of $20 million.

THE INVESTIGATION

Within a month after the Alviso residents were allowed to return to their homes we were asked by an attorney if we were available to provide consultation as to the causes of the flooding. He represented the law firms that had been jointly retained by many of the residents in an attempt to secure restitution for their losses. After meeting with the attorney several facts were immediately obvious:

1. It was so soon after the event that the attorneys needed very basic information to proceed. At that point they knew very little about the March 1983 flooding and its historical roots or the roles of the government agencies involved.
2. The attorneys, because they were working on a contingency fee arrangement, were not anxious to expend a great deal on consulting until they knew they had a reasonably winnable case against someone.

This was the ideal situation for us to propose a concise preliminary investigation and report for a modest lump-sum fee. If the findings were such that a lawsuit seemed plausible, they either could authorize us to continue our forensic work or could drop us and use other consulting firms, depending on the exact nature of the expertise required. If their interpretation of our investigation was that litigation was not warranted, they could either so inform their clients or employ others who might develop new concepts or uncover data more favorable to the instigation of a lawsuit.

The attorney was pleased with the proposed preliminary report approach.

We began the investigation with little or no specific knowledge of the Alviso flood and certainly no preconceived notions or theories. The preliminary investigation consumed about 2 mo, most of the time spent in the field and gathering data from local government agencies. Even though there had been no legal actions filed, we found most government officials reluctant to discuss the subject freely. However, much information was contained in files that were open to public scrutiny. Local newspapers were extremely helpful too.

Many of the residents were interviewed, not only in regard to this particular event but relative to the historical aspects of the problem. Several of the residents had lived in Alviso long enough to have spanned three generations of public servants. Outlined here are the agencies we contacted and the material sought.

Principal Sources of Information
1. City of San Jose
   (a) Planning Department
   (b) Building Department
2. Santa Clara Valley Water District
3. USGS
4. *San Jose Mercury* and other local newspapers
5. U.S. Army Corps of Engineers, San Francisco Office
6. Federal Emergency Management Agency, San Francisco Office
7. State Coastal Commission (San Francisco Bay Resource Conservation District)

Nature of Information Obtained
1. Aerial photographs
2. Aerial topographic maps
3. Flood insurance maps
4. Reports
   (a) Subsidence
   (b) Flood control recommendations
   (c) Historical
   (d) Flood insurance
5. Building permit data
6. Newspaper accounts
7. Correspondence and memoranda reviewed in water district library
8. Precipitation data
9. Hydrologic and hydraulic data for Coyote Creek
10. Tidal data

While the results of our preliminary investigation were not conclusive and would require extensive detailed analyses and verification, we presented the following as the main causes of the flooding:

1. The extreme subsidence of the Alviso area was caused by excessive groundwater extraction. While the entire valley floor had subsided, the

Alviso area was more critically affected because it was naturally at or only slightly above sea level.

2. Because of high antecedent rainfall and the nature and operation of the dams, no storage was available in Anderson and Coyote Reservoirs at the time of peak flow.

3. High antecedent rainfall had caused saturation of watershed soils, rendering them unable to absorb additional precipitation; hence greater runoff occurred.

4. A combination of peak flow and high tide caused the water-surface elevation to rise to elevations higher than the lowest tops of the Coyote Creek levees and the temporary ring levee.

5. Once water topped the levees it was trapped within Alviso and had to be pumped out.

6. Lower Coyote Creek was poorly maintained.

7. Coyote Creek levees were inadequate to contain a 10-yr storm.

8. Several impediments to free surface flow existed, including the City of San Jose wastewater treatment plant and the levees constructed for salt harvesting in the bay.

We found that there were three public agencies with primary control or responsibility for conditions related to the cause and effect of the flooding: the Santa Clara Valley Water District, the City of San Jose, and the U.S. Army Corps of Engineers.

Santa Clara Valley Water District
1. Operated most of the wells that pumped the groundwater and lowered the aquifer
2. Set the storage and discharge policy for and regulated the operations of the Anderson and Coyote Reservoirs and Dams
3. Was responsible for construction, operation, and maintenance of the major flood control facilities, including the channels and levees

City of San Jose
1. Constructed the wastewater treatment plant and its protective levees
2. Issued building permits in Alviso
3. Generally responsibile for protection of Alviso citizens and property including operation and maintenance of the floodgates and ring levee

U.S. Army Corps of Engineers
1. Generally responsible for approval of activities within "navigable waters" of the bay including the issuance of permits for salt harvesting

Our preliminary investigation laid the foundations for our attorney clients to prepare a case against the Santa Clara Valley Water District, the City of San Jose, and the salt company operating in the bay.

For the next 2 yr we did very little additional investigation, most of our activity being confined to assisting the law firms in preparing interrogatories and helping with other discovery functions. The legal effort appeared hampered and delayed by frequent changes in personnel, each change requiring fresh briefings and countless meetings.

In early 1987 the final team of lawyers came aboard and forcefully took charge. Time was drawing short to prepare for trial, and we were authorized, almost carte blanche, to perform all the additional studies we felt were essential to complete our investigation. In assessing the work to be done and the time available to do it, we suggested to the attorneys that they augment the consulting team by employing a civil engineering firm with state-of-the-art expertise and computer capability in complex hydrologic analyses. An internationally renowned consulting firm possessing such attributes was quickly employed. Most of the ensuing engineering investigation described herein was ably and professionally performed by that company.

Areal Subsidence

The withdrawal of groundwater and its relationship to subsidence was well documented in a series of reports prepared by the USGS under its principal author, J.F. Poland. Much of the information and data under this heading have been extracted and condensed from these reports. Figure 9–3 depicts the underlying geology in the Santa Clara Valley.

Most of the subsidence in the San Jose area of the Santa Clara Valley occurred over a 50-yr period from 1916 to 1966. It was a period marked by below-normal rainfall coupled with a nearly 400% increase in groundwater withdrawal. In response, the artesian head dropped 180–220 ft (54–66 m). As a direct result of the artesian head decline the land surface subsided as much as 12.7 ft (3.8 m) in San Jose owing to the compaction of the fine-grained compressible confining beds and interbeds as their pore pressure decreased. Subsidence was brought under control by 1966 through greatly expanding the importation of water and, at the same time, increasing groundwater recharge.

To test the compressibility of the underlying soil strata, the USGS drilled a series of core holes in 1960 across the floor of the valley. An electric log was obtained for each core hole after the coring was completed. Graphic logs and generalized lithologic descriptions were prepared from the geologists' logs made at the drill sites, supplemented by interpretation of the electric logs in zones not cored or of poor recovery. These elements were combined to give a composite log for each core hole.

Samples from these core holes were laboratory tested for physical and hydrologic properties including particle size distribution, specific gravity of solids, dry unit weight, porosity and void ratio, hydraulic conductivity (normal and parallel to bedding), and Atterberg limits.

Compressibility characteristics of fine-grained compressible layers were obtained by making one-dimensional consolidation tests of undisturbed

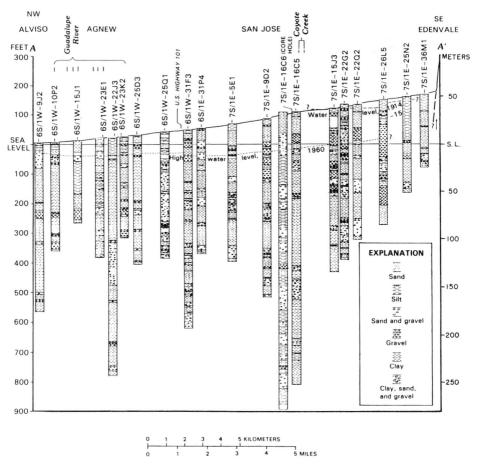

FIGURE 9–3

Geologic cross section (from Poland and Ireland, 1988).

samples. Properties tested included the compression index, a measure of the nonlinear compressibility, and the coefficient of consolidation, a measure of the time rate of consolidation. The combined results of these tests for each coring were then compared with actual ground-surface subsidence in the test core areas as measured by survey leveling data. Correlation was generally close, proving that the subsidence was due to soil compression as the result of groundwater extraction.

The most important factor in the recovery of artesian head was the importation of surface water from two sources:

1. The Hetch Hetchy Reservoir in the Sierra Nevada through a purchase agreement with the City of San Francisco, owner of the reservoir
2. The State of California South Bay Aqueduct, which transports water from the Central Valley

See Figure 9–4 for import volumes.

The magnitude of the subsidence that occurred in north Santa Clara County from 1934 to 1967 is shown in Figure 9–5. The subsidence contour running through Alviso is 6 ft (1.8 m). Early USGS maps indicate that Alviso was at an average elevation of 2 ft (0.6 m) above sea level. After the effects of subsidence it was about 4 ft (1.2 m) below sea level. Figure 9–6 shows the relationship of subsidence to groundwater levels and population.

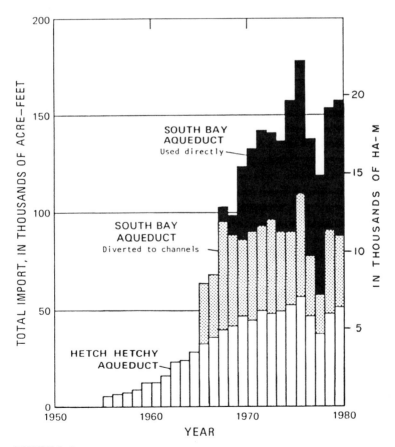

FIGURE 9–4

Surface-water imports to north Santa Clara County, 1955–1980 (from Poland and Ireland, 1988).

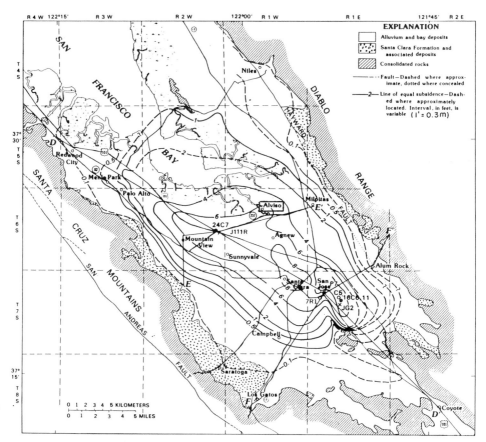

FIGURE 9–5
Subsidence contour map (from Poland and Ireland, 1988).

Coyote Creek

The inability of the lower reaches of the Coyote Creek channel and levee system to contain the floodwaters constituted a basic causation factor in the 1983 Alviso flood levels.

Coyote Creek and its tributaries drain a total of 366 mi^2 (952 km^2), representing about 25% of all of Santa Clara County. Two dams and reservoirs, Coyote and Anderson, were constructed in 1930 and 1950, respectively, along the 75-mi (120-km) Coyote Creek watercourse. The primary purpose of the reservoirs was for water conservation, with flood control a secondary function.

Over half of the Coyote Creek watershed lies upstream of Anderson Dam,

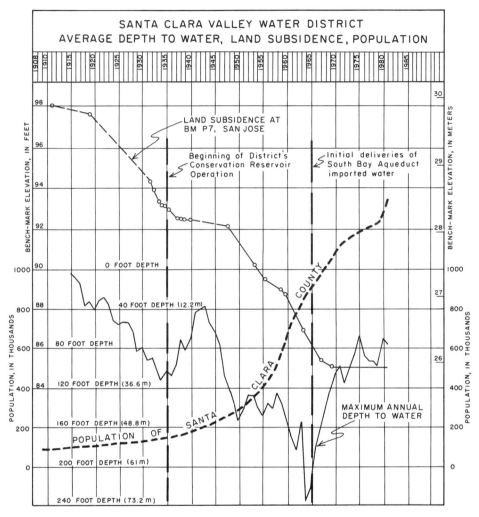

FIGURE 9–6

Relationship between depth to water, land subsidence, and population growth.

where the terrain is mountainous and the precipitation greater than in the valley. The highest elevation in the watershed is in excess of 3000 ft (900 m).

In the 40-mi (64 km) downstream section between Anderson Dam and San Francisco Bay, the hydraulic capacity of Coyote Creek rapidly diminished (as of 1983) at:

Metcalf Road	25,500 ft³/sec	(765 m³/sec)
Highway 17	9,500 ft³/sec	(285 m³/sec)
Dixon Road	2,000 ft³/sec	(60 m³/sec)

Reasons for this dramatic decrease in capacity include

1. Natural flattening of the longitudinal streambed gradient reduces the velocity of flow and hence the capacity of the channel. (In theory, if all other factors remained constant, the velocity would decrease in proportion to the square root of the gradient.) The velocity reduction from the upper to the lower reaches of the 40-mi (64 km) segment was in the range of 15 ft/sec (4.5 m/sec) to 1 ft/sec (0.3 m/sec).
2. Channel improvements in the lower reaches between U.S. Highway 237 and the bay have been minimal. Levees that existed in this segment were grossly inadequate in both height and structural integrity. Most were nonengineered, but rather built and maintained by adjacent farmers to protect their crops. The Federal Emergency Management Agency (FEMA) does not recognize nonengineered levees in determining federal flood insurance rates.
3. Because of the extremely low velocities at the terminus of the channel during flood flows, a substantial amount of siltation takes place, thereby reducing capacity.
4. The land subsidence accentuated the tidal effects, decreasing the hydraulic gradient and the capacity of the channel. Because of the subsidence, the levee systems in this area are perched, that is, the tops of the levees are (or should be) higher than highest tide elevations.

While Coyote Creek south of Highway 237 was the pre-1983 recipient of many beneficial flood control improvements, the segment to the north was largely ignored. Since financing priority for such improvements was predicated mainly on the value of the adjoining real estate, there apparently was little money available for the Alviso area.

As the result of all these capacity-reducing elements, Coyote Creek was incapable of containing a flood of a 10-yr return frequency.

Rainfall Statistics

From the standpoint of total seasonal rainfall, the 1982-83 season produced 30.25 in. (76.8 cm) as contrasted to an annual average of 14.34 in. (36.4 cm). It represented the highest seasonal total since 30.30 in. (77.0 cm) fell in 1889–1890. See Figure 9–7.

While not having a direct relationship to maximum flood flow, the abnormally high seasonal total contributed to the flooding in Alviso in at least two ways:

1. It kept the reservoirs full during the critical period, permitting no storage.
2. It saturated the surface soils within the Coyote Creek watershed, producing a higher coefficient of runoff.

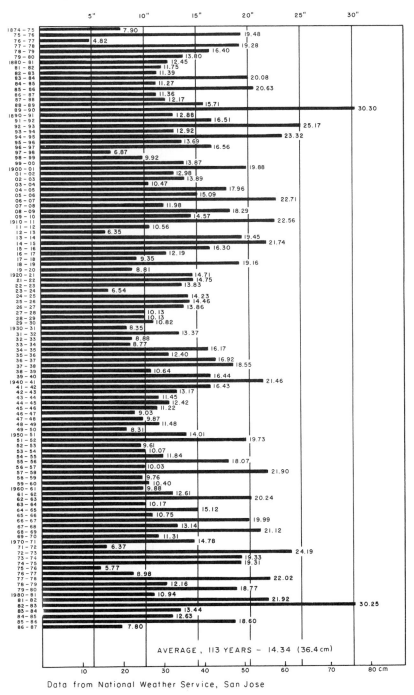

FIGURE 9–7

Seasonal rainfall totals at San Jose airport.

Flood Magnitude

In the typical forensic investigation of a flood it is important to determine the peak flow (Q) within the watercourse causing the flooding. There is generally a direct correlation between the Q and the maximum depth of the flood-waters. Also, the Q can be compared with historical peak flow data to determine the approximate return frequency (as once in 5,10,50, 100 or more years) of the subject flow. The government entity having responsibility for flood control has usually already done the statistical work and has formulated return frequency data, which make for easy comparison to the subject Q. Knowing the approximate return frequency of the Q serves to put the event in the proper perspective relative to its predictability. Since most flood control projects are designed for a minimum 100-yr return frequency, it is difficult to argue that a discharge of a lesser return frequency is unpredictable and therefore constitutes an act of God.

The peak Q in lower Coyote Creek during the March 1983 flood was 5600 ft^3/sec (168 m^3/sec). It occurred on March 2 and was measured by an automatic recording stream gage. Statistically the Q represented a return frequency of about 10. For comparison:

25-yr $Q =$	8,500 ft^3/sec	(255 m^3/sec)
50-yr $Q =$	12,000 ft^3/sec	(360 m^3/sec)
100-yr $Q =$	14,500 ft^3/sec	(435 m^3/sec)

The time of peak flow appeared to slightly precede the time of the highest flood levels in Alviso. Generally flood levels subside rather quickly after peak flow has occurred. However, in Alviso the outflow-restricting effect of the makeshift ring levee, the downstream constructions, the high tide, and the inability of the discharge pumps to keep up with the continuing inflow caused the flood levels to rise even higher, as well as to subside much more slowly than normal.

Generally flooding damage to structures can be attributed to four factors:

1. Maximum high water
2. Duration of inundation
3. Velocity of flood flow (deposition of silt with low velocities and erosive damage with high velocities)
4. Existence of debris in flood flow

In the case of Alviso, damage to the residences was made more severe by the artificially heightened water levels and the prolonged period of high water accompanied by a corresponding greater volume of silt and other sediments deposited as the result of low velocities.

Tide Gates

During the 1940s 16 tide gates were constructed along the east bank of the Guadalupe River, at the Marina, and at other locations north and east of Alviso. They were operated by a "floodgate master" in Alviso, who opened them at low tide to allow accumulated floodwaters to drain out to the bay and closed them at high tide to prevent back-up of seawater into Alviso.

The system worked well until Alviso was annexed to San Jose in 1969. Then the tide gates were apparently not maintained, fell into a state of disrepair, and were eventually removed or abandoned.

Removal of 10 − 4 × 8 ft (1.2 × 2.4-m) tide gates from the east bank of the Guadalupe River and the subsequent raising of this levee in 1963 and 1983 created a barrier that trapped floodwaters and caused a greater depth of flooding than otherwise would have occurred in the 1983 flood. These 10 tide gates were used in the past to allow any floodwaters that may have accumulated on the floodplain between Coyote Creek and the Guadalupe River to exit into the Guadalupe before they reached Alviso. They were an important and obvious drainage feature on the floodplain near Alviso and were removed in 1963 when the Guadalupe River was channelized. No equivalent device was installed to allow drainage of water before it reached Alviso. During the flood of March 1983 the Guadalupe River levee acted as a barrier, causing floodwaters to pond in Alviso.

Hydrologic calculations indicated that if the tide gates had been operational in March 1983, the peak flood level in Alviso would have been reduced by about 6 in. (15 cm). However, their primary benefit would have been to drastically reduce the duration of the submerged condition, thereby lessening damage and allowing the residents to return to their homes much sooner.

Reservoir Operation

The water district has operated the Coyote and Anderson reservoirs on Coyote Creek primarily for water conservation and secondarily for flood control. Capacities of the two reservoirs were:

Coyote	23,700 acre ft	(2,915 ha m)
Anderson	91,300 acre ft	(11,230 ha m)
Total	115,000 acre ft	(14,145 ha m)

Anderson Reservoir was designed to store runoff from its watershed (including overflow from upstream Coyote Reservoir) and to release it slowly through an outlet structure in the dam for recharge of the groundwater through percolation as well as for irrigation purposes. The flood control capability of the system derives from the storage volume available during peak storm periods and to a lesser extent from a degree of attenuation provided even with full reservoirs.

Water can be released through the outlet structure at a rate of 500 ft³/sec (15 m³/sec) to make room for more runoff. However, to release water during the rainfall season, when little of it can be absorbed in the saturated recharge basin and when there is no irrigation demand is wasteful. Moreover, in periods of heavy prolonged rainfall there is the danger of the inflow rate exceeding 500 ft³/sec (15 m³/sec) for a long enough period to fill the reservoir and cause water to overtop the spillway of the dam. These periods can produce discharges in excess of the capacity of the downstream Coyote Creek channel and levee system, particularly in the Alviso area. Although overtopping of Anderson Dam has only occurred five times in its 50-yr history, the consequences for Alviso have been damaging on four of those occasions. In the subject 1983 flood, overtopping took place for 67 consecutive days.

Recognizing the need for a rational approach to reservoir level control, the water district developed a "rule curve." Empirically derived through use of historical rainfall and runoff data for the Anderson Reservoir watershed, the objective of the rule curve was simply to achieve a theoretic full reservoir each April at the end of the rainfall season. The rule curve shows the theoretic desired volume of water in the reservoir at any given time. Since it represents a statistical goal it does not change from year to year except when subject to revisions in historical rainfall data. The rule curve for Anderson and Coyote Reservoirs is shown in Figure 9–8.

The "operating curve" shown for the 1983 storm was entirely arbitrary. It represented what the water district thought was a more practical approach, one that would result in greater water storage. A very tangible economic gain is to be realized in maximizing storage and recharge. Every acre foot stored represents one less acre foot that has to be purchased through the California State Water Project.

Under both the rule curve and the operating curve, storage is gradually decreased from April to November for recharge and irrigation until it reaches a level of approximately 67,000 acre ft (8241 ha m), about 73% of capacity. Then when actual storage exceeds the curve storage, controlled discharge up to 500 ft³/sec (15 m³/sec) occurs until actual storage coincides with the curve being used.

In 1982 and 1983 the operating curve was used for release of water rather than the rule curve. After the 1982 floods, in which water spilled over the dam between March 25 and April 17, the water district followed the routine of gradually releasing water until approximately 65,000 acre ft (7995 ha m) remained in Anderson Reservoir in November. Two storms occurred in December that prompted controlled releases in order to adhere to the operating curve. However, a severe storm in late January 1983 filled Anderson Reservoir to capacity because of a faster rate of inflow than of controlled

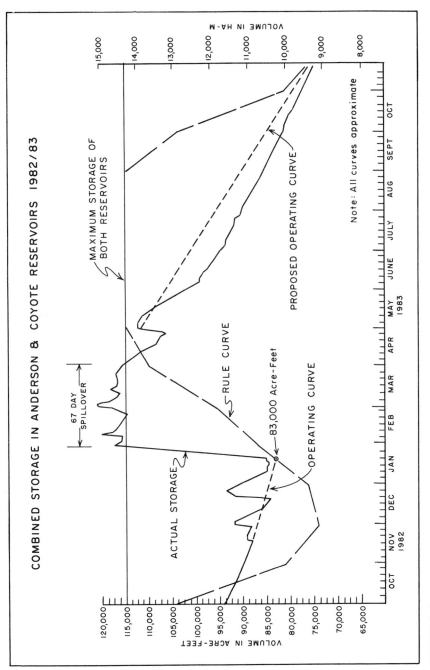

FIGURE 9–8
Rule curve for Anderson and Coyote Reservoirs.

discharge. The problem was exacerbated by the fact that, because of fear of damage to the outlet structure, discharge rates were not allowed to exceed 350–400 ft³/sec (10.5–12.0 m³/sec). Consequently water spilled over the dam for 67 continuous days as shown in Figure 9–8.

The maximum rate of flow in the lower reaches of Coyote Creek during the overtopping period was approximately 5600 ft³/sec (168 m³/sec). It occurred on March 2, the day that Alviso was inundated, and can be seen in the actual storage curve peak.

The investigation and analysis of the discharge procedure that were actually followed, as they related to the rule curve and the operating curve (both officially adopted by the district's board of directors), revealed the following:

1. The valve that controlled the outlet structure discharge was not functioning properly on January 11, when it was almost completely closed. The discharge rate on this date was only 40 ft³/sec (1.2 m³/sec).

2. From January 18 to January 20 the valve was not opened even though 6 in. (15.2 cm) of rain fell in the watershed during the period and the actual storage curve was higher than the operating curve.

3. If the valve had been discharging at its design rate of 500 ft³/sec (15 m³/sec) instead of an average of 375 ft³/sec (11.2 m³/sec), spillover would have been decreased about 250 acre ft (30.8 ha m) or 81 million gal (307 million liter) per day during the critical period.

4. If the total combined storage of the two reservoirs had been 60,000 acre ft (7380 ha m), 52% of total combined capacity in November, there would have been no overtopping and, hence, no flooding in Alviso.

5. If the rule curve had been followed there would still have been flood control storage on March 1. Although overtopping would have occurred after that date, its peak magnitude would have been much less than that which occurred on March 2 and probably would have been contained within the downstream levee system. Even if not, flood damage would have been minimal.

Wastewater Treatment Plant

Construction of the San Jose/Santa Clara wastewater treatment plant, including the accompanying sludge ponds (shown in Fig. 9–2), contributed significantly to the Alviso flood damage. The $40 million facility was built in 1962 and underwent considerable expansion in the mid-1970s. At the time of the 1983 flood it occupied a total area of about 1200 acres (480 ha). Both the plant and the sludge ponds were protected from flooding and tidal intrusion by levees 10–12 ft (3–3.6 m) high designed to withstand a 100-yr flood flow.

One of the adverse impacts of the plant was that its levees diverted the natural flow of floodwaters. Water that would have otherwise sheet flowed across the site of the facility and into the Artesian Slough was diverted westerly toward Alviso. Since the salt company levees had already created a major obstruction to natural flow, the treatment plant sealed off one of the few remaining drainage courses to San Francisco Bay.

Prior to construction of the treatment plant, Artesian Slough had constituted a major drainage path to the bay for water overflowing Coyote Creek. With its construction, the function of the Artesian Slough was converted to that of an outlet for treated wastewater. The slough was channelized on both sides and sealed from any legitimate hydraulic connection to the Alviso floodplain area. In addition, the Artesian Slough watercourse south of the treatment plant was filled in for the construction of a railroad spur line and replaced with a 36-in. (91.4-cm) culvert pipe. The culvert, however, was regulated by a gate and pump and thus could drain only a small quantity of water. During a flood period, Artesian Slough is usually filled with imported water that enters the sewage system from other parts of the county. Thus the gate arrangement on the culvert pipe is not functional. Water ponded on both sides of the Artesian Slough levees by March 2, 1983, indicating that flood-waters were being blocked from their natural drainage course.

Eventually the levees were breached, but this was too late to prevent the flooding of Alviso. San Jose had been put on notice concerning the Artesian Slough blockage in a 1974 letter from a professor at a local major university. The letter stated that the sludge-pond construction should provide for fresh-water flooding to go into Artesian Slough.

The four sludge ponds cover approximately 800 acres (320 ha), representing more than one-fourth of the total floodplain area north of Highway 237. Construction of the ponds displaced water on the floodplain, causing an increase in flooding depth as well as creating a flow obstruction. Despite the obvious adverse impact, no mitigation measures were taken to protect Alviso from the increase in flooding caused by the construction of these ponds. It is the practice of FEMA to require such mitigation when a development project causes more than 1 ft (30 cm) rise in the floodplain elevation.

Construction of the treatment plant's sludge ponds involved raising the height of an S-shaped levee, shown in Figure9–2, which blocked floodwaters from flowing into the Dead Arm of Coyote Creek (also known as Mud Slough), a former natural drainage path. The previous levee had an elevation of 7 ft (2.1 m). According to a report by a city employee, the S levee was raised by the City of San Jose to about 10 ft (3 m) supposedly to protect the sludge ponds from tidal flows. The S-shaped levee was eventually breached during the March 1983 flood because water was ponding next to it. The capacity of the Dead Arm of Coyote Creek is substantial and, had the S levee been in its original condition, the quantity of floodwater reaching the Alviso area would have been minimal. A hydrologist expert witness employed by the City of San Jose was in agreement. He stated in his deposition that the maximum water-surface elevation near Alviso would have been 0.7 ft (0.2 m) had the S levee been at its original elevation of 7 ft (2.1 m). This is 4.6 ft (1.4 m) less than the actual average maximum flood elevation of 5.3 ft (1.6 m). Since most of the streets of Alviso have an elevation greater than 1 ft (0.3 m), significant flood damage would not have occurred.

Salt Harvesting

Over the years most of the marshland at the south end of San Francisco Bay has been converted to salt evaporation ponds, a strictly private commercial venture under permit from the U.S. Army Corps of Engineers. A major expansion of the operation occurred in the 1950s wherein the salt company constructed and reconstructed an extensive system of levees encompassing 4000 or 5000 acres (1600 or 2000 ha) of ponds.

The levees, designed to block high tides from entering the ponds, also drastically restricted floodplain drainage from reaching the bay. With the construction of the salt-pond levees, flow was generally confined to the Guadalupe River and Coyote Creek itself. When those watercourses overflowed or breached their levees, floodwaters were blocked by the salt-pond levees, precipitating a rise in the Alviso flood levels. This was definitely the case during the 1983 flood.

Urbanization

Between 1950 and 1980 the population of Santa Clara County more than quadrupled. Unfortunately the floodplain management functions essential to accommodate this growth did not keep pace.

Urbanization of the valley dramatically increased storm runoff for three fundamental reasons:

1. Increased runoff coefficients
2. Decreased "time of concentration" (t_c)
3. Reduced storage areas

As the valley was transformed from agricultural to residential, commercial, and industrial land uses, the rainfall runoff coefficient increased sharply. With a typical farm or orchard in this region, about 15% of the precipitation in a major storm could be expected to run off into the streams and rivers. The remainder would either infiltrate to the subsurface or be consumed in evapotranspiration. Under identical storm conditions after development, the typical runoff coefficient percentages would be on the order of:

Single-family residential	45%
Multiple residential	55%

| Commercial | 75% |
| Industrial | 85% |

The maximum rate of runoff from any watershed can be roughly expressed in the basic Rational formula, $Q = ACI$, in which Q is the peak discharge in cubic feet per second, A is the area of the watershed in acres, C is the coefficient of runoff, and I is the average rainfall intensity in inches per hour during a period of time that corresponds to the time of concentration (the time required for rain falling at the most remote point in the watershed to reach the discharge point).

Obviously, based on the runoff coefficient only, development can increase peak discharge by three to six times. Development also markedly decreases the time of concentration since flow velocity is much greater on paved surfaces and in storm drain pipes than on underdeveloped or cultivated land. Development of areas that were previously natural ponds or sumps also decreases the time of concentration and increases the peak downstream runoff.

Influence of High Tides

Further research on this subject revealed that the high tide in San Francisco Bay apparently reached a maximum of 9.1 ft (2.7 m) on both March 1 and March 2. These tides were not abnormally high in relation to those predicted for February and March. A maximum high tide for 1983 of 10.6 ft (3.2 m) occurred in January.

The fact that tides at the time of peak flow were not higher than normal does not mean that they did not have a profound influence on the flooding levels in Alviso. It only indicates that the high tides that occurred on March 1 and 2 could not be singled out as a rare act of God causation factor.

There was no evidence that tidal levels were in any way raised by lowered atmospheric pressure or seismic activity.

River Landing Road Bridge

In the early 1970s a bridge was built over Coyote Creek at River Landing Road by the owners of a solid waste disposal facility located northeasterly of the wastewater treatment plant. The narrow bridge opening and low bridge deck created an obstruction to high-volume flows in Coyote Creek. However, analysis of the actual March 1983 flood flow indicated that the bridge contributed only slightly to the overbank flow because channel capacity was so low upstream of the bridge.

Investigation Conclusions

It was obvious to us that the 1983 Alviso flood damage was caused by a combination of events and circumstances—physical, administrative, and political—that produced a cumulative impact.

To estimate the relative amount of flooding contributed by each of the individual physical factors, a numerical flood routing model was developed by our team consulting firm. The model was based on a computational technique known as continuity routing and was calibrated to reproduce the approximate timing and water-surface elevations that occurred during the flood event. The March 1983 flood hydrograph was used as the upstream boundary condition, with a dynamic tidal water-surface elevation in the bay as the downstream boundary condition. The sequence and timing of specific events were pieced together from news accounts, eyewitness testimony, high-water marks, and public documents describing the flood. After the model was calibrated it was used to evaluate the approximate contribution of individual physical factors by examining the relative degree of flooding (changes in predicted flood elevation in Alviso) for several hypothetical combinations of physical factors.

Results from the modeling studies indicate the approximate contribution to the flood by the following individual physical factors:

- Operation of the reservoirs and upstream channel mainte-   15%
  nance by 1982
- Effects of the 12-ft (3.6-m) S levee   15%
- Construction of the system of salt evaporation ponds   10%
- Lack of flood control planning and provision of flood control 5%
  drainage facilities in Alviso
- Absence of the former tide gates   10%
- Effects of the wastewater treatment plant, including removal 25%
  of flood storage volume, creation of a barrier to flow, and
  altering historic drainage channels
- Land-surface subsidence   20%

While this computer relative causation analysis was interesting, it might have been difficult to explain its extensive and complex programming to a judge or jury had there been a trial. It seems sufficient to state that all the physical factors outlined were substantial causes of the flooding.

The principal administrative and political causation factors (all related to the physical) could be summarized as follows:

- Failure to follow the officially adopted policy in operating the reservoirs
- Failure to provide Coyote Creek flood control improvements commensurate with urban growth
- Failure to provide an engineered ring levee for Alviso until after the event

- Failure to mitigate the effects of the wastewater treatment facility
- Failure to mitigate the effects of the salt pond levees
- Decision to remove or abandon the tide gates and replace their function with an emergency pumping program
- Failure to provide adequate maintenance of Coyote Creek downstream of Highway 17

THE LITIGATION

Attorneys for the plaintiffs brought suit against the City of San Jose and the Santa Clara Valley Water District alleging breach of mandatory duty, dangerous condition of public property, common-law negligence, and inverse condemnation. The salt company also was sued for not taking affirmative actions to prevent or ameliorate the Alviso flooding. The total amount of damage sought by the plaintiffs in these suits was $40 million, including amounts for pain and suffering as well as physical damage.

The defendants employed cadres of legal and consulting help. It was exceedingly important that they win the case, not only because of the high dollor amount involved but also because of the precedent-setting implications of a loss, including the possible need to revamp their entire floodplain management policies and priorities.

Pretrial legal maneuvering was a protracted and expensive process. We got the distinct impression that the "deep pockets" defendants were deliberately attempting to impose an unacceptably high financial burden on the plaintiffs' lawyers in the hope that they would quietly settle out of court for a more modest sum. Depositions of the expert witnesses not only were lengthy but many were very bitter and abusive. One of the plaintiffs' witnesses, a young but extremely well qualified hydrologist, was grilled unmercifully for two straight weeks, with much of the time spent impugning his lack of seniority in the profession. It was the first deposition of his career, and as he remarked afterwards, "Maybe my last." It was not exactly a pleasant initiation to the legal rites of forensic engineering.

Just before trial all the defendants made a motion for summary judgment or, alternately, summary adjudication of certain issues. The judge granted summary judgment to the salt company, thereby removing one defendant from the case. He stated in his decision that there was no dispute as to what the defendant did and did not do before and during the 1983 flood, only whether the salt company had a duty to the plaintiffs to do something in addition. His ruling was that they did not. Unlike a precedent case in which the defendants had erected a barrier that diverted floodwaters on their neighbors, he ruled the salt company took no action that actually diverted floodwaters.

The judge also ruled for summary adjudication for the city and the water district, dismissing all causes of actions and allegations *except that of inverse condemnation.*

The plaintiffs' attorneys planned to appeal both the summary judgment and the summary adjudications, but before they could do so the judge ordered a series of settlement conferences. After much debate and negotiation a settlement was reached that paid the plaintiffs a total of $12 million. Of course, this figure included their legal fees.

We could only speculate as to why the defendants decided to settle. Four reasons came to mind immediately:

1. They could easily have lost the case on inverse condemnation as well as the appeal on the summary judgment.
2. None of the defendants wanted what would be mainly adverse publicity in connection with the trial.
3. A hydrologist expert witness for the City of San Jose had rendered an opinion in a deposition that if the S levee had not been rebuilt to protect the sewage treatment plant, the flooding in Alviso would not have exceeded 0.7 ft (0.2 m).
4. Because it had provided funding to the City of San Jose for flood control improvements, and was reportedly investigating the management of the funds, FEMA took more than a passing interest in this case. The Department of Justice not only monitored the depositions but eventually employed its own consultants to perform an independent hydrodynamic study of the flooding at Alviso. The fact that this study generally confirmed our analyses probably made the defendants even more wary of a trial.

As far as the plaintiffs' attorneys were concerned, their reasons for accepting the settlement were more obvious. At that point they felt that the settlement amount was sufficient to warrant risking a complete loss to both their clients and themselves.

THE POST MORTEM

A Case of Benign Neglect

When the judge upheld, in part, the defendants' motion for summary judgment, he dismissed all causes of action and allegations of the plaintiffs related to breach of mandatory duty, dangerous conditions of public property, and common-law negligence. This adjudication was based largely on his interpretation of case law. Whether his decision would have been sustained on appeal is, of course, a matter of conjecture.

It is difficult for our nonlegal minds to reconcile all of the actions and omissions—physical, administrative, and political—that were detrimental to the citizens of Alviso without assessing a measure of negligence, however benign, to the defendants. It appeared that the only reason for this neglect was the comparatively low value of Alviso real estate. Two preeminent reflections arise from the judge's decision of nonnegligence:

1. The courts seem to assess less liability to a series of smaller, diverse cumulative acts of commission or omission than they do to one large act, even though the resulting damage may be identical.
2. Use of strict cost–benefit formulae to prioritize either new construction or maintenance work often do not take into account the relative liability exposure resulting from delaying a project or assigning it a lower priority. A prime example is that, after the Alviso damage had been done, San Jose was able to quickly find $50,000 to build a more substantial ring levee.

Tide Gates versus Pumps

Of the many errors in judgment that contributed to this disaster, the decision by San Jose to abandon the tide gates in favor of pumping the water out of Alviso ranked near the top.

For many years the tide gates, by their very simplicity, had played a major role in mitigating the flood problem in Alviso. They were merely opened to provide drainage from Alviso when the tide was low and closed to prevent backup of San Francisco Bay waters when the tide was high. Because they were so simple to operate and maintain, they represented a cost-effective method of partial flood control.

Despite the high tides that were coincidental with the floodwaters, there were still periods of low tide when the tide gates could have discharged large volumes of water from Alviso. Given all the other factors contributing to the flooding, the pumps would still have been needed. However, if the floodgates had been functioning, the high water level in Alviso would have been reduced and the duration of the flooding decreased considerably, thereby significantly reducing the damage.

What Is "Diversion"?

In upholding the salt company's motion for summary judgment, the judge, in effect, ruled that the company's salt-pond levees did not "divert" flood-waters into Alviso. This was a decision based on case law regarding drainage diversions. Not being attorneys and not being familiar with all the appropriate case law on the subject, we are certainly in no position to question the legal reasoning that led to his decision.

From the standpoint of a hydrologic/hydraulics design specialist, there are two basic situations to avoid assiduously. The first is the diversion of drainage flow from the natural path whether that flow is confined to a defined watercourse or is in the form of sheetflow. When diversion occurs outside a floodplain it is usually easy to recognize. Because the flow is generally supercritical, control of flow comes from upstream, and most downstream activity can have no effect. The second situation to avoid, as within an estuarial floodplain, is where diversion is generally much less identifiable, where the flow is subcritical, and where control is downstream. Such is the case of Alviso, where downstream obstructions to flow reflect upstream in the form of higher flood elevations.

Whether such downstream blockage of natural floodplain flow falls within a legal definition of diversion is unimportant from an engineering point of view. The flood damage is just as avoidable as with upstream diversion in a non-floodplain area. Floodplain management generally reflects a much higher level of understanding and sophistication than it did 20 yr ago. FEMA has played a leadership role, not only in mapping for flood insurance purposes but in establishing design criteria for agencies and engineers to use in the planning and design of structures and facilities within floodplains. One of the design criteria advocated by FEMA and adopted by most floodplain management agencies is that construction within a floodplain is not permissible if it causes the 100-yr flood level of any other property within that floodplain to be raised more than 1 ft (0.3 m). While the 1-ft (0.3-m) dimension is arbitrary and the hydrologic calculations often very complex, the rule has proven to be a workable and effective measure of control.

If one were to apply this rule to Alviso, the salt company would have been in violation. However, two arguments could probably be made for the salt company:

1. It would be unfair to make such a rule retroactive
2. The salt ponds are not really in the floodplain proper but in San Francisco Bay. (It is our understanding that the Corps of Engineers had jurisdiction in "navigable waters." Then comes the question of how one defines navigable waters, ad infinitum.)

To recap, the judge may have made the correct legal decision in terms of the case law he considered. If that is true there needs to be a great deal more work done toward expanding the legal definition of diversion.

The Rule Curve

This subject was discussed briefly in Chapter 6 wherein Mexican officials were faced with a similar problem of balancing the need to conserve valuable water in the Rodriguez Reservoir against the peril of downstream flooding if they conserved too much.

There was no question that the Anderson and Coyote Dams and Reservoirs were planned and designed primarily for water storage. However, the fact that a rule curve was established is indicative of the concern for flood control too. Two questions emerge relative to Anderson Reservoir's flood control role:

1. Why wasn't the adopted rule curve followed in the 1982–1983 season?
2. Why didn't the original design of the reservoir and dam incorporate greater capacity for flood control? (If the dam had been 5 ft [1.5 m] higher, it would have resulted in enough additional storage to have averted the flooding in Alviso. The additional cost at that time would have been nominal. Plans were prepared after the 1987 flood to raise the dam 5 ft [1.5 m], but the work has never been done—probably because of the high cost to retrofit.)

Legal Costs

The concept of employing mitigation in lieu of litigation did not stand a chance in this case. After enduring decades of unmitigated exposure to flood hazard, the residents of Alviso were in no mood to barter dropping the lawsuit for promises of mitigative action. Besides, the inadequacy of flood control had deteriorated to the point where the city and the water district were probably not in the financial position of being able to guarantee protection. However, one has to wonder how far the total estimated legal fees of $6 million would have gone toward providing a measure of protection to Alviso.

Present State of Protection

Eight years after the flood it is interesting to examine what steps have been taken to reduce the likelihood of a repeat performance:

1. The ring levee built immediately after the flood is still in place and apparently adequately maintained. However, the ring levee remains a stopgap measure, as structurally reliable as its weakest link, and is probably not high enough for protection against a 25-yr storm superimposed on a full reservoir. Even if not breached or overtopped, the ring levee isolates the community until the water level in the surrounding area has subsided.
2. The Corps of Engineers has been placing test fills in preparation for construction of new levees along Coyote Creek from San Francisco Bay to Highway 237. Construction will begin at the downstream edge and work its way upstream, which is in sharp contrast to historical improvements of Coyote Creek. When completed the downstream reach of Coyote Creek will provide 100-yr flood protection.

Duty of Government

Enlightened floodplain management may eventually succeed in precluding new construction in floodways that are not properly protected. In the interim, government has an obligation to protect those structures already built under legal permit or built before permits were required. It was particularly disturbing to find that permits for new buildings in Alviso had been issued by the City of San Jose after 1969.

Granted the perpetual scarcity of funds for this purpose, efforts should be made at every political level to find methods of financing. Even setting aside the human toll exacted by flooding, the threat of costly litigation is reason enough to assign a high priority to flood projects. Low-cost federally subsidized flood insurance has been a huge help, but structural solutions are the only long-range answer to the problem.

In contrast to most of the disasters portrayed in previous chapters, this one was not a singular episode lacking serious portents. Rather it was the culmination of a series of less devastating floods with a steady increase of potential consequences due to ever-increasing growth and development within the watersheds without a corresponding increase in protection.

Keeping Abreast of Urbanization

As individual subdivisions and commercial projects were developed in the Santa Clara Valley, each had a very small effect on peak discharge downstream. However, in the aggregate they had a profound impact. This compounding effect was apparently never properly translated into concrete physical and financial planning for the floodplain. While it would have been politically unpopular, each residential subdivision and commercial development should have been forced, as a permit requirement, to pay its pro rata share of the cost of providing downstream flood protection. Such assessments would have been a fair and equitable means of allocating the cost of flood control facilities and would have provided a permanent interest-bearing account from which funds could have been drawn for particular projects. Maintenance money has traditionally come from specific or general tax revenue and financing of capital improvements from county-wide bond issue elections, which are not popular ways to raise money. It seems to make much more sense to assess the people who are specifically increasing the storm discharge. The developers, in turn, would have passed these costs on to the buyers of the homes and commercial entities. This method of financing not only would have been more equitable than general taxation but would more likely assure the funds being available when they were needed.

A recent appellate court decision in San Diego County overruled a trial court and found in favor of plaintiff downstream property owners whose properties had been flooded as the result (or at least partly so) of upstream development. The findings of the appellate court could be summarized as follows:

- An upstream landowner has no absolute right to protect his land from floodwaters by constructing flood control structures that increase the downstream flow of water in its natural watercourse, but is instead governed by the ordinary principles of negligence.
- Local government must consider downstream consequences and take action to prevent damage.
- If the upstream owner modifies the stream, he must consider downstream consequences.

The second finding was applicable to the City of San Diego. It stated, in effect, that the city should not have permitted upstream development without providing (or requiring the developer to provide) compensating mitigation for its downstream effects.

10 RIVER MORPHOLOGY: MINING VERSUS AGRICULTURE

THE EVENT

Northern California's Russian River is home to some of the most lush fruit orchards and vineyards in the world. Flanked by rolling foothills and blessed with fertile soils and a climate tempered by its proximity to the Pacific Ocean, the river basin is ideally suited for the growing of plums, pears, apples, and grapes.

The Russian River was first settled in 1812 by a colony of Russian immigrants who established a fort at the mouth of the river near what is now Guerneville (Fig. 10–1). At first the Russians were chiefly preoccupied with hunting sea otter in collaboration with native Indians. However, they quickly recognized the river basin's agricultural potential and began growing a variety of crops.

Because of its mild climate, serene beauty, and relatively close proximity of 75 mi (120 km) to the metropolitan San Francisco Bay region, the area became a magnet for well-heeled bay area residents who bought land, converted it to orchards, built homes, and, with hired help, ended up with income-producing weekend retreats. That is not to say that all the orchards belong to absentee part-time farmers. Most are owned and operated full time by families who have retained their properties for generations. It is a rare heir who would sell a piece of this bucolic agrarian paradise. All this desirability has, of course, made the land quite valuable—too valuable to be eroded and washed into the sea.

Beginning in the early 1970s riparian farmers along the middle reach of the Russian River (a 9-mi [14.4-km] stretch between Healdsburg and the downstream Wohler Bridge) and the lower portion of its principal tributary named Dry Creek noticed that segments of the riverbanks were eroding, steepening, and collapsing. In some instances structures built close to the banks were being threatened if not actually damaged.

This phenomenon was not new to the old-timers, who had witnessed some degree of bank erosion in every major storm. In most instances erosion had

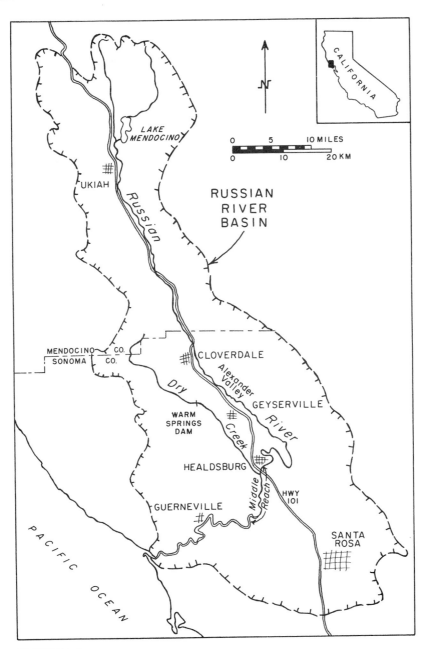

FIGURE 10–1
Vicinity map of the Russian River Basin.

Typical orchard along Dry Creek. Note evidence of instability on opposite riverbank.

been mitigated by the farmers through the use of simple, nonengineered protection such as ungrouted riprap or pipe and wire revetments. What made the 1970s unique was the much greater magnitude of the erosion and bank failure and the inability to stop or even slow the process with conventional measures.

It was observed that the streambed itself was degrading, causing over-steepening of the banks and their ultimate collapse. The rivers were becoming wider and deeper at the expense of the adjacent land. Between 1970 and 1978 it was estimated that over 70 acres (28 ha) of prime farmland had been lost along the middle reach and Dry Creek owing to this phenomenon.

Dating back to the late 1800s, the Russian River and its tributaries, particularly Dry Creek, have been a source of high-quality aggregate for the construction industry. Sand and gravel from the riverbeds as well as from hillside quarries were supplied not only locally but to the entire region. Many large-scale construction projects such as Hamilton Air Force Base and the Golden Gate and Richmond/San Rafael Bridges used aggregate mined from the Russian River watershed. The need for aggregate paralleled the rapid regional population growth. By 1978 the annual demand was in excess of 4 million ton (4.06 million metric ton). To meet this damand, private companies were mining 17 sites along the Russian River and Dry Creek (see Fig. 10–2).

There was no question as to the importance of the gravel industry to the

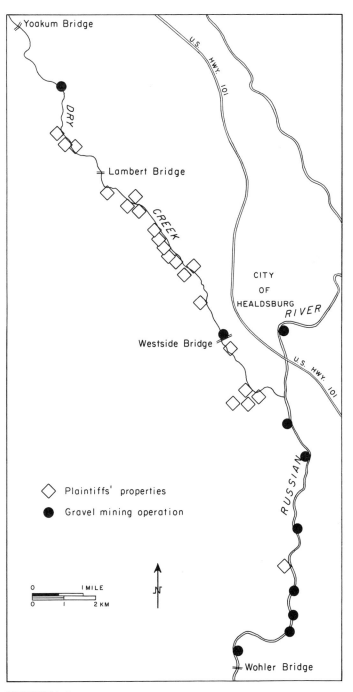

FIGURE 10–2
Map of Dry Creek and the middle reach of the Russian River.

Sonoma County economy. The gravel companies estimated that 20% of county payrolls were linked directly or indirectly to the industry.

Three basic sources of gravel have been available to operators in Sonoma County:

	Approximate relative volumes as of 1980 (%)
In-stream	10
Terrace	80
Hard-rock hillside quarries	10

In-stream mining was done within the stream channel, generally by means of drag lines. Large pits were excavated to depths of about 60 ft (18 m) below the water level or, if found, shallower layers of clay. The pits were naturally filled during periods of high flow with material that contained not only sand and gravel but large quantities of clay and silt.

Terrace mining was conducted adjacent to the stream channel by excavating similar pits. However, terrace pits remained open until filled either by an extremely high overbank flood flow or by reclamation for agricultural use.

It was estimated that there were about fifty hard-rock quarry sites available in the county—enough to provide all the construction industry's needs for over 50 yr. The quality characteristics of the rock were identical to those in the streams but, because of the additional cost of blasting and transportation, the overall cost of extraction and marketing was higher than that of river mining.

It was not until about 1970 that suspicions surfaced linking in-stream gravel mining to upstream riverbank instability and the resulting loss of farmland. To attempt to determine the cause or causes of the problem, a noted expert in river morphology was asked by Sonoma County officials to make a preliminary investigation. His tentative conclusion was that the extreme streambed degradations and consequent bank failures were probably due to excessive downstream gravel mining in the years before 1970. This hypothesis was also expounded by the State of California Division of Mines and Geology. The lengthy time gap between alleged cause and effect was explained by the very slow upstream migration of the streambed degradation known as the "knick-point."

THE INVESTIGATION

In 1978 we were approached regarding our forensic consulting availability by an attorney representing 22 of the Russian River and Dry Creek riparian property owners. He informed us, in general terms, of the riverbank instability problems suffered by the farmers as well as the scope of the gravel mining operations.

Before committing our services we accepted an invitation for a private flight to the area to make a cursory inspection of the most obvious evidence of recent streambed degradation. Extreme downcutting was particularly apparent in the vicinity of the Westside Bridge, which spans Dry Creek near Healdsburg. The bridge was located between two of the largest in-stream gravel operations. Erosion around the bridge abutments had been so severe as to nearly expose the foundations. Riparian vegetation that normally occurs at the water line was perched about 10 ft (3 m) above the stream surface. New growth at the degraded stream level was of small stature, indicating that the downcutting was of relatively recent vintage. Upstream from the bridge there was man-made evidence of the recent downcutting in the form of riprap, used for bank erosion control, which was perched up to 8 ft (2.4 m) above the surface of the creek. As we walked upstream we noticed a progressively gradual decrease in streambed degradation.

That initial field trip convinced us that the streambed erosion we saw was

Westside Bridge over Dry Creek at Healdsburg. Extreme streambed degradation that was due, in the authors' opinion, to excessive in-stream extraction of sand and gravel downstream from bridge caused exposure of bridge pier footings.

not the result of normal stream processes and that it would be necessary to look for abnormal contributing causes such as:

1. Unusually high incidence of erosion-producing major storm flows
2. Recent fire–flood events within the upstream watershed
3. Land-use changes within the upstream watershed that might have significantly increased peak runoff discharges
4. In-stream activities by the property owners such as construction of small dams, revetments, and levees that might have disturbed natural flow characteristics
5. Activities that might have upset natural bed-load erosion, transport, and deposition processes such as in-stream gravel mining

Since streambank and streambed erosion is normally a function of stream-flow velocity (hence discharge), it was essential to examine the peak flows that occurred between 1940 and 1978, especially in Dry Creek, where most of the streambank failure occurred. Stream gage records were obtained from the U.S. Army Corps of Engineers, which showed the following peak discharges (Q) in Dry Creek at the Yoakum Bridge:

Date of storm	Peak discharge (Q)	
	(ft³/sec)	(m³/sec)
01/21/43	28,000	840
12/03/50	23,800	714
12/06/52	20,300	609
01/17/54	23,000	690
12/22/55	31,600	948
02/24/58	27,900	837
02/08/60	20,400	612
01/13/62	25,100	753
01/31/63	32,400	972
12/22/64	31,800	954
01/23/70	27,700	831
01/12/73	22,300	669
01/16/74	32,000	960

In 1978 there were also several high-intensity storms in the watershed. However, by 1978 the upstream Warm Springs Dam construction had attenuated the peak flows in Dry Creek. To put these discharges into context we calculated the return frequency of a 10-yr Q to be about 32,000 ft³/sec (960 m³/sec) and of a 5-yr Q about 24,000 ft³/sec (720 m³/sec).

An examination of the peak discharges indicated that the 38-year period produced a normal number of 5- and 10-yr flows and nothing above a 10-yr. The fact that there was no discharge greater than that of a 10-yr return frequency in the 1970s was instrumental in our eliminating unusually high velocities as a primary cause of the excessive bank failure and land loss. It also seemed particularly meaningful that the two consecutive 10-yr discharges in 1963 and 1964 did not result in the excessive erosion that occurred in the 1970s when two similar Qs occurred in 1973 and 1974.

We could find no physical evidence or records to indicate that any significant wildfires occurred in the watershed before the 1973–1974 storms.

As far as land-use changes were concerned, there had been steady population growth in the watershed with accompanying residential and commercial development. But since the stream discharges shown previously were based on actual stream gage measurements, changes in runoff characteristics had already been accounted for.

We physically inspected many of the riparian properties that had been damaged, as well as interviewed the owners, to determine if individually constructed in-stream erosion control facilities had played a part in the degradation process. Though it was impossible to protect one riparian property without exerting a small degree of hydraulic influence on others, we concluded that any adverse effects had been extremely minor.

Having eliminated all other factors as being improbable or insignificant contributors to streambank failure, we focused our attention on the effects of the in-stream sand and gravel mining operations. Our investigation of the impact of gravel extraction necessitated answers to the following basic questions:

- How much in-stream sand and gravel was removed on an annual basis and where was it removed?
- What time frame should be used for the study?
- What changes occurred in the streambed profiles and cross sections during this period?
  (1) In the river reaches where the plaintiffs' properties were located?
  (2) Elsewhere?
- What was the natural replenishment rate of bed-load material during the chosen time frame?

Extraction Quantities

Quantities, locations, and, in some cases, depths of sand and gravel extractions were obtained from the gravel companies' own records secured through legal process and through the records of the California Division of Mines and Geology. This information was analyzed and integrated on a year-to-year basis starting in 1946, the first year of recorded large-scale operations in the middle reach. A summary of the analysis of all in-stream records indicated the following approximate extraction quantities:

	Period from 1946 through 1972		Period from 1973 through 1978		Total	
	(Million ton)	(Million metric ton)	(Million ton)	(Million metric ton)	(Million ton)	(Million metric ton)
Middle reach of Russian River	27.5	27.940	1.2	1.219	28.7	29.159
Dry Creek	1.0	1.016	1.2	1.219	2.2	2.235
Total	28.5	28.956	2.4	2.438	30.9	31.394

Time Element

We decided to frame our investigation between the years 1940 and 1978. The first comprehensive topographic mapping of the rivers was done by the Corps of Engineers in 1940. It provided a reasonable correlation to the topography that existed in 1946, when large-scale mining began and formal extraction records were kept. Undoubtedly some mining occurred in the intervening World War II period but, as far as we could determine, was not of sufficient magnitude to materially affect our studies.

Stream Changes, 1940–1978

Topographic data in the form of field survey notes, cross sections, and profiles were obtained from the San Francisco District of the U.S. Army Corps of Engineers, the Sonoma County Water Agency, the U.S. Department of Agriculture Soil Conservation Service, and the U.S. Coast and Geodetic Survey. The Corps of Engineers performed comprehensive surveys in 1964 and 1972 as well as 1940, while the water agency surveyed specific critical areas in 1978. Using these data we were able to prepare streambed profiles and bank-to-bank cross sections for both the Russian River and Dry Creek as follows:

1. **Russian River** From the mouth of the River at the ocean to a point about 64 mi (102 km) upstream at Preston Bridge north of Cloverdale. Basically we were able to study changes in streambed elevations and bank-to-bank cross sections that occurred along this entire 64-mi (102-km) reach from 1940 to 1978.
2. **Dry Creek** From the confluence with the Russian River to a point about 14 mi (22 km) upstream at the Warm Springs Dam site. Because of all the studies and planning prepared in conjunction with the Warm Springs project, the Corps of Engineers was able to supply us with comprehensive survey data. From these data we were able to study in considerable detail the changes in streambed elevations and bank-to-bank cross sections that occurred along this 14-mi (22-km) reach from 1940 to 1978.

In addition to topographic data, the Corps of Engineers and the Sonoma County Water Agency were able to provide us with a wealth of high-clarity aerial photographs comprehensively covering both the rivers and the watershed. Dates of the photographs spanned the entire period of our investigation, enabling us to identify incremental changes. Scales varied, permitting a wide range of studies from general geology and development within the watershed to mining operations to stream erosion and bed-load movement. The importance of quality aerial photography in environmental-type forensic investigations cannot be overemphasized.

To supplement the survey and aerial photographic data, several field trips enabled us to more accurately evaluate soil types, temporary erosion control measures, age and type of riparian vegetation, rock outcroppings, and man-made structures such as dams and levees. All available reports in connection with streambed degradation, bank erosion, sediment transport, bed-load material replenishments, rainfall and runoff records, and other related technical matters were carefully evaluated.

Figure 10–3 depicts the 1940 and 1972 profiles of the thalweg (streambed) of the middle reach of the Russian River. Also included are several 1978 spot elevations showing further downcutting between 1972 and 1978 at certain critical locations. The maximum degradation in the middle reach was over 19 ft (5.7 m), with the average about 11 ft (3.3 m).

Similar thalweg profiles were plotted for Dry Creek as well as for the lower

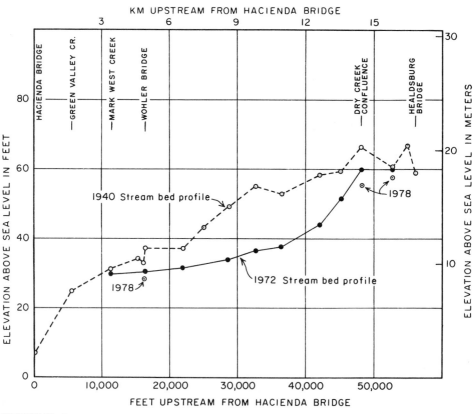

FIGURE 10–3

Profile of the middle reach of the Russian River.

and upper reaches of the Russian River. At its Russian River confluence, Dry Creek had degraded 11 ft (3.3 m). This general magnitude of degradation continued upstream about 2 mi (3.2 km) to the Mill Creek tributary, from where it gradually diminished until a condition of equilibrium was reached near the Lambert Bridge (about 7 mi [11 km] upstream from the confluence). Profiles along the upper and lower reaches of the Russian River indicated that there had been very slight net degradation, with many locations showing aggradation.

Natural Replenishment

All streams transport sediment eroded from their watersheds as well as from their own sides and beds. The volume of sediment transported varies with the intensity and duration of the rainfall, the geology of the watershed, and the condition of the watershed relative to recent wildfire, earthwork and grading activity, mining, etc.

Transported sediment is categorized as suspended load or bed load depending on its weight or volume and the velocity of flow. The smaller particles in suspension generally constitute the bulk of the material carried. The suspended load may travel by saltation, which is partly by suspension and partly by traction along the bed. Particles carried by saltation are described to progress by a series of leaps and rolls. A large portion of the suspended load is not carried mechanically but is dissolved in the water.

The bed load is composed of the larger particles—sand, gravel, and larger rocks—which slide, roll, or bounce along the streambed. We were concerned only with the bed load in evaluating aggregate replenishment capability, since the smaller particles are not mined.

We could find no evidence that sampling and monitoring of the bed-load transport had ever been accomplished. Without benefit of such a program it would have been impossible to accurately estimate average annual replenishment. However, in terms of our assignment, a geometric approach could establish the net effect of the gravel mining. Assuming the middle reach and Dry Creek as parts of the same regimen:

$$A = B + C - D$$

where
- A = Total bed-load replenishment from both the Dry Creek watershed and the upstream Russian River
- B = Total bed-load material extracted
- C = Total bed-load material passing through the middle reach
- D = Net volume lost because of degradation of both riverbeds

B had already been calculated as a total of about 31 million ton (31.5 million metric ton). C was difficult to estimate but was of minimal quantity. Considering the depth and number of pits, it seemed inconceivable that the multiple gravel operations permitted substantial volumes to pass through. For the purpose of our analysis we assumed the figure to be about 10% of the mined volume, or 3 million ton(3.05 million metric ton). D was calculated by plotting periodic cross sections of both rivers based on Corps of Engineers topography for 1940 and 1972 supplemented by 1978 Sonoma County Water Agency topographic data. Using simple average-end-area calculations, we estimated the volume lost between 1940 and 1978 to be:

Middle reach	11,400,000 yd³	(9,120,000 m³)
Dry Creek	4,600,000 yd³	(3,680,000 m³)
Total	16,000,000 yd³	(12,800,000 m³)

or the equivalent of about 24 million ton (24.38 million metric ton) of bed-load material.

Thus $A = 31,000,000 + 3,000,000 - 24,000,000 = 10$ million ton (10.16 million metric ton) total replenishment from 1946 to 1978, or an annual average of about *312,000 ton* (317,000 metric ton).

To roughly corroborate this figure we estimated bed-load replenishment based on a 1971 USGS report, "Turbidity and Suspended Sediment Transport in the Russian River Basin." The authors (Ritter and Brown, 1971) found the suspended sediment yield in the Dry Creek watershed of 217 mi² (564 km²) to be 5,770 ton/mi² (15,240 metric ton/km²) per year. This figure was predicated on an average of data compiled over a 4-yr period from 1964–1965 to 1968–1969. Use of this period may have resulted in a slightly higher than average annual yield owing to the occurrence of a 10-yr return frequency storm.

Text research indicated that the ratio of bed-load material to suspended material, taking into account specific watershed characteristics related to geology, topography, and rainfall, was about 1 : 10. On this basis, the average annual bed-load eroded and transported would have been about 600 ton/mi² (1580 metric tons/km²), or a total of about 130,000 ton (132,00 metric ton) per year. Applying the same 600 ton/mi² (1580 metric ton/km²) per year figure to the upstream approximate 500-mi² (1300-km²) contributing Russian River basin (below Coyote Dam) yielded an estimated bed-load transport to the middle reach of 300,000 ton (305,000 metric ton) per year. The total of 430,000 ton (437,000 metric ton) per year was 27% higher than the 312,000 ton (317,000 metric ton) in the geometric analysis. However, even this simplistic check gave us confirmation that our geometric approach was not unreasonable.

Velocity Consideration

The presence of the large gravel pits in the streambeds created steeper thalweg or streambed gradients and, hence, higher velocities upstream of the excavations. Higher velocities, in turn, resulted in greater erosive forces, which manifested themselves in the following ways:

1. In the flow of the water itself:
   (a) Direct erosion of the streambanks
   (b) Decrease in side pressure adjacent to banks contributing to their collapse
   (c) Hydrodynamic uplift on the stream bottom contributing to down-cutting
2. In the force of the particles moved by the water. The particles were picked up by the stream flow in three ways:
   (a) Impact
   (b) Frictional drag
   (c) Hydrodynamic uplift

Empirical studies have indicated that the weight or volume of the particles that can be moved varies with the sixth power of the velocity. So with a spherical particle the radius varies with the square of the velocity. Obviously a small increase in velocity can result in a relatively large increase in the size of the particles carried, which is reflected in greater erosional force.

To assess the effect of the gravel pits on upstream velocities we analyzed a reach of Dry Creek 6900 ft (2070 m) long between points 1.1 mi (1.8 km) and 2.4 mi (3.5 km) upstream of the Westside Bridge. This reach was chosen because it contained eight litigant properties and had suffered the highest concentration of land lost through streambank failures (about 42 acres [17 ha]).

Using the 1964 and 1978 topographic survey data, we calculated the average thalweg gradient in this reach to have changed from 0.0012 in 1964 to 0.0021 in 1978. Since the velocity varies with the square root of the gradient in the Manning formula

$$V = \frac{1.486}{n} R^{2/3} S^{1/2}$$

the velocity would have increased about 30%, which is enough to have had a major effect on erosional forces. Of course this assumes that all other components of the formula remained constant, which they did not. However, changes in the stream cross sections resulted in a greater hydraulic radius for high flows with about the same roughness coefficient, thereby further increasing the velocity.

Warm Springs Dam

The U.S. Army Corps of Engineers designed and built the Warm Springs Dam and reservoir (Sonoma Lake) on Dry Creek at a site about 11 mi (18 km) upstream from Healdsburg. The purpose of the project was threefold: flood protection, water supply, and recreation. Although some preliminary relocation construction began in the early 1970s, the stream itself was not disturbed until 1978. Therefore the dam had no effect on our bed-load replenishment studies relative to the litigation. However, for the recommendations we made to the Sonoma County Board of Supervisors relative to a proposed aggregate management plan, the effects of the dam on future bed-load replenishment were vital.

Investigation Summary

In analyzing the changes that had occurred in the 64 mi (102 km) of the Russian River streambed profiles, we found that the approximately 9 mi- (14-km-) long middle reach was the only section that experienced excessive degradation and erosion. It was also the only reach where in-stream gravel extractions other than selective gravel-bar removal had occurred. From 1940 to 1978 the net lowering of the streambed in the middle reach exceeded 19 ft (5.7 m) at one point and averaged about 11 ft (3.3 m) over the entire 9 mi (14 km). We calculated the net volume of bed material lost during this period to be about 10 million ton (10.6 metric ton). *At no other reach, either upstream or downstream of the middle reach, had this excessive degradation occurred.* In fact the profiles upstream in the Alexander Valley indicated a general process of both mild buildup and mild lowering, not exceeding 5 ft (1.5 m) anywhere, for the same time periods. This was true despite the fact that all other conditions that contributed to degradation and erosion were present to a comparable degree in the Alexander Valley reach. These comparable factors existed over the entire 64-mi (102-km) stretch and included stream alignment, stream gradient, stream velocities, soil conditions, watershed development, watershed fires, and flood conditions.

The Dry Creek profiles and cross sections also indicated that excessive degradation and erosion primarily occurred in the lower section, including the Mill Creek tributary, where the effects of the overall imbalance between extraction and replenishment were felt. From a point about 1 mi (1.6 km) upstream from Lambert Bridge to the Warm Springs Dam site, the streambed and banks appear to have been relatively stable. Again, all the other factors contributing to degradation and erosion were present to a comparable degree upstream as well as downstream.

In attempting to reach a condition of equilibrium, the degraded streambed began a downcutting or "notching" action, whereby a knickpoint migrated upstream from the middle reach up Dry Creek (and Mill Creek) until it joined the degraded section at the Westside Bridge. From this point it then proceeded upstream to beyond Lambert Bridge, where it eventually reached a point of equilibrium.

The downcutting had two basic detrimental effects on adjacent riparian property. First, in many cases the increased height of streambanks caused the banks to collapse. Second, the increase in streambed gradient increased stream-flow velocities during high-discharge storms. Since even a slight increase in velocity has a profound effect on erosive force, both the banks and the bed were subject to much greater erosive actions. Also, the gravel pits created turbulence, which in itself has a tremendously erosive effect. All these factors combined to cause the excessive erosion and damage to the land of certain riparian property owners.

Public Hearings

This investigation was unique from the standpoint of how its results were revealed. With other forensic assignments, in which litigation or potential litigation was present, our investigative work was done in an atmosphere of strict confidentiality. Findings were not ordinarily disclosed until we were deposed and, even then, only to the extent required by the specific questions asked. Some of the results of our other investigations were not revealed until the time of our direct trial testimony.

In this case a political inquiry proceeded along a parallel path with the litigation. The Sonoma County Board of Supervisors conducted a lengthy series of public hearings to consider all options of an aggregate management plan prepared by the county's planning staff. One of the key elements of the plan was to place an immediate ban on in-stream gravel mining in Dry Creek and the middle reach of the Russian River.

Naturally our clients were campaigning vigorously for the adoption of the plan. The gravel operators, just as naturally, were in fierce opposition. Because of the far-reaching economic and environmental impacts at stake, there was widespread and highly vocal representation at the hearings from both sides of the issue.

Even among the riparian farmers there was no unanimity. While those upstream of the extraction sites strongly favored the proposed plan, many of those whose properties abutted the sites benefited from the flood protection provided by the increased hydraulic capacity of the channel. Conversely, farmers located immediately downstream of the extraction site were subjected to varying degrees of increased bank erosion because of higher flood flow velocities emanating from the excavated channels.

In short, the proposed plan had a wide range of supporters and opponents, some of who were influenced by personal considerations and others who were concerned with broader economic or environmental issues.

Both sides in the lawsuit decided to have their consultants appear at the public hearings to offer their data and opinions as well as to present written reports. The hearings, in effect, became a harbinger of the trial, with a notable difference: The hearings were based on the potential future adverse impacts of in-stream gravel extraction, whereas the trial would focus on past alleged adverse effects. Despite this distinction, and that one decision would be political and the other legal, both sides fully realized the long-range implications of the hearings and the decision by the board of supervisors. Consequently no data were held back at the hearings to save for the trial.

In July 1980 we submitted a report to the board of supervisors that summarized our investigation and conclusions to date. By that time there was no question in our minds that the imbalance between gravel extractions and replenishment was the cause of the extreme streambed degradation and consequential streambank instability. We therefore made the following recommendations to the county relative to future aggregate management in the middle reach and Dry Creek:

1. Establish an effective sampling and monitoring program to balance extraction and replenishment and to prevent future instability and excessive erosion. This is vitally important because Warm Springs Dam will cause a drastic reduction in bed-load material transport for two reasons. First, all material eroded in the stream and in the watershed upstream from the dam will be trapped by the dam. Second, the dam will also cause a dramatic decrease in discharges and velocities during storms of peak magnitude. It is during such storms that most of the material is transported. For example, the decrease in discharge for a 10-yr storm is from about 30,000 ft³/sec (900 m³/sec) before construction of the dam to about 7,000 ft³/sec (210 m³/sec) after completion.

2. Temporarily prohibit all in-stream mining, other than selective removal of gravel bars that could cause lateral erosion, until the system has had an opportunity to aggrade to a more stable condition and to ascertain the effect of the Warm Springs Dam on bed-load material transport in Dry Creek.

3. Prohibit permanently all in-stream ''pit'' mining and restrict removal to material above basic streambed grade.

4. Reevaluate and reconsider the potential for hard-rock sources considering that:

    (a) The in-stream gravel has been derived by natural erosion processes from those same rock materials.

    (b) Quarry sites will have to be located where hard, durable bedrock exists (and such sites do exist according to geotechnic consultants in Sonoma County).

    (c) The quarry sites can be reclaimed for housing and other uses provided eventual use is considered during extraction.

The consultants to the gravel companies submitted a great deal of material to the county in support of their opinion that to shut down or force drastic reductions of in-stream extraction was not warranted. It was their position that previous estimates (including ours) of bed-load transport and, hence, replenishment rates were more than 300% low and that the subject required considerably more study before any conclusions could be reached. Their explanations for the obvious imbalance was that there had undoubtedly been a large volume of extractions in the early years (before 1946) by unlicensed operators that had gone unreported. This postulation, if it could have been proven, would have served the gravel companies well in two respects. First, it might have been a persuasive argument in support of going slowly in banning in-stream mining. Second, it might have been used as an argument in trial to dilute the alleged impact of the legitimate gravel operations. To our knowledge no attempt was ever made to quantify the volume of unlicensed gravel operations.

In trying to reconcile the huge difference in estimated replenishment rates between ourselves and the gravel companies, we found that their experts may have included all sand material including the fine-grained portion in suspension as part of the replenishment material. If so, this was not appropriate since only the more coarse-grained sand (0.15 mm or larger) would have been part of the bed load and have had commercial value.

Whether to adopt the proposed aggregate management plan posed a difficult dilemma to the board of supervisors. The gravel industry constituted a large segment of the county's economy and the board could ill afford to saddle the industry with unnecessarily restrictive conditions. However, after months of hearings and much deliberation, the board voted unanimously in July 1980 to adopt the proposed ordinance.

THE LITIGATION

Based on our early opinion that the in-stream gravel operations constituted the proximate cause of damage to certain riparian properties, a lawsuit was filed by the property owners against the major gravel companies as well as the Sonoma County. The county was named as a defendant because it had also extracted gravel from Dry Creek.

The gravel operators, in turn, cross-complained against the property owners, alleging they had constructed individual makeshift erosion control improvements within the stream, which caused damage to neighboring properties. Depositions were taken of all the experts on both sides. In our case we made the mistake of agreeing to being deposed before our investigation had been completed. In-depth questioning by several astute gravel company attorneys, one of whom was also a civil engineer, made us realize that we still had a lot of work to do. However, before further depositions were scheduled, the public hearings had become our forum for release of data and opinions, making more depositions redundant.

While the board of supervisors was engaged in the final throes of public hearings on the proposed gravel management plan, an out-of-court settlement was reached between the gravel operators and the association of property owners. Under the terms of the settlement, four of the largest gravel companies agreed to pay a net total of $705,000 to the 22 litigants plus several other riparian landowners. Included in the overall settlement were several insurance company payments, including $50,000 to one of the gravel operations in the cross-suit. The county's insurance carrier also paid $25,000. The settlement conference lasted 2 wk, the longest in the history of Sonoma County litigation.

Besides the cash settlements, the gravel companies yielded some vested rights (grandfather clauses in their licensing agreements), including the right to extract gravel from the Dry Creek streambed below the low water line.

The settlement did not acknowledge admission of responsibility on the part of the gravel companies for the damage caused the farmers. However, the gravel operators did not appear anxious to endure a protracted trial, which had been projected to last 6 mo to 1 yr with corresponding astronomic legal fees. In addition the public hearings on the county's aggregate management plan were winding down, with the eventual vote appearing to be adoption of the plan. From a long-term public relations standpoint, it seemed important for the gravel industry to settle the lawsuit.

For the plaintiff farmers the settlement brought mixed emotions. Some were not satisfied with their individual damage payments, but, on the whole, they did not want to risk everything on an unpredictable court verdict a year away. Their in-hand cash payments plus the likely prospect of adoption of the management plan for future gravel operations made a settlement acceptable.

This case seemed to us another example of legal overkill. It was estimated that there were 35 law firms representing plaintiffs, defendants, and insurance companies. Legal documents filed in the lawsuit filled over 30 folders in Superior Court records. On the other hand, physical mitigation of past damages would not have been feasible. Mitigation of future damage was reasonably assured with the adoption of the county's aggregate management plan.

THE POST MORTEM

A Classic Trade-off

The time has long passed when one industry or one interest can wantonly exploit a valuable public resource at the expense of another legitimate interest or of the community at large without inviting a legal response. On the other hand it is often possible for conflicting interests to reach a compromise position that will be to everyone's advantage in the long run. The Russian River conflict appeared to be one of these cases.

Gravel extraction from a riverbed is unlike most other types of mining in that the product has been recently (in a geologic time frame) transported by natural forces to the excavation site from upstream property owned by others. In a sense, extracting gravel might be thought of as being in the quasi-public domain. If one accepts this thesis, the gravel operators, even if they own the land or have secured legal excavation rights, have an obligation to mine the gravel in a manner that would minimize adverse effects on riparian landowners and on the environment in general.

If large-scale gravel operations in the middle reach of the Russian River had been begun in 1980 instead of 1950, a comprehensive environmental impact report (EIR) with its attendant public hearings, etc. would have been required. Those who obtained their mining rights prior to the advent of EIRs should have no less responsibility to their neighbors and to the public.

There seems little controversy that in-stream gravel mining in the middle reach can safely continue if careful monitoring is done to ensure a long-term balance between extraction and replenishment. The future of terrace mining is a difficult subject involving broad consideration of aesthetics, reclamation, and ultimate land use. Hard-rock or hillside mining will have to increase to make up for the decrease in in-stream volumes as well as the gradual depletion of terrace deposits. Since the basic rock characteristics are similar, hard-rock mining should not result in decreased quality of the aggregate. Quantities are plentiful, but gravel prices will increase as the result of increased operational and transport costs as well as site reclamation and other expenses necessary to mitigate environmentally adverse factors such as noise, dust, and visual impact.

Profile of Equilibrium

Every river or stream reaches its "profile of equilibrium" when its slope and volume are in equilibrium with the sediment load it transports (referenced to the average-sized material on its bed). It is a temporary condition, constantly changing as the velocity and volume of flow change, sometimes degrading and sometimes aggrading. However, the long-range tendency is generally toward more erosion than aggradation. Hence one could expect most streams to gradually deepen with corresponding lateral corrasion.

In the absence of human interference with this natural striving toward balance, episodes of flooding, erosion, and aggradation and the detrimental effects they may have on riparian properties are accepted as natural events for which no one can be blamed. When human activity interferes with the natural profile of equilibrium and someone is hurt as a result, a lawsuit will generally follow. It does not matter that other property owners may have benefited from the activity. We are not aware of any legal process whereby those who benefit pay those who are damaged. Major human interference with river flow such as dams are generally built by government entities after considerable input by all parties who will be directly affected as well as by environmental organizations. At least in principle the decision to build a dam is based on an overall assessment of benefits versus costs and liabilities. As government is being stripped of liability immunity and with the ever-broadening interpretation of inverse condemnation, dams are not being taken lightly relative to potential litigation. In the case of privately operated river activities such as gravel mining, it is absolutely essential that they be thoroughly analyzed in advance to determine possible adverse impacts and then closely monitored to be certain the operation is proceeding according to plan.

11 AN OVERVIEW

MAINTAINING A PROFESSIONAL APPROACH

If there is a single overriding tenet in this book it is this: Engineers and geologists who are employed as forensic experts owe it to themselves, to their professions, and to society as a whole to be as objective and professional as is humanly possible. It is regrettable that the words "as is humanly possible" had to be included. But we are convinced that, despite the best and most determined intentions, sometimes it is humanly impossible for some to remain aloof and detached, given the highly competitive adversarial nature of the legal process and the demands of the attorney for whom the expert is working.

Because the opportunities for forensic engineering and geologic work have expanded so rapidly, there has been little time for the participants to step back and evaluate what they are really doing and what impact their work is having on their personal reputations and on their professions.

Of all the professionals engaged as expert witnesses, the ones who have probably received the most media exposure are the psychiatrists who testify in court as to the sanity or state of mind of a criminally charged defendant. In a highly publicized case there may be as many as three or four psychiatrists, all with outstanding credentials, testifying avidly for each side. The spectacle inevitably arouses certain questions:

1. Is the science of psychiatry so imprecise that six experts could honestly have such diverse opinions?
2. Did these experts accept their assignments before making a preliminary investigation to determine if they could live comfortably with their commitments?
3. Did they accept the assignments with the knowledge that they are sufficiently adroit to be able to conjure up scientific theories to justify any opinion?
4. Are not the two teams of experts really interchangeable?

The public reaction to this situation seems generally apathetic. The view that most psychiatrists who testify in court are hired guns who sell themselves to the highest bidder seems to have been accepted as part and parcel of the criminal justice system. This acceptance may be due to the relatively inexact nature of psychiatric diagnosis.

It is our observation that the engineering and earth science professions are falling into similar disrepute on the issue of expert testimony. Blatant advocacy is rampant—almost out of control. We see so many prominent civil engineers and geologists who were thoroughly objective and professional throughout the early portions of their careers, when they were engaged in their conventional roles as planners, designers, construction managers, and administrators, suddenly abandon their high professional standards when they became involved in forensic work.

It may be that many feel there is nothing wrong with being an advocate. Certainly nearly all experts become advocates to some degree. It seems impossible, in a fiercely competitive environment, not to have an interest in being a part of the winning team. However, for engineers and geologists we feel there is a line to be drawn, to wit:

- Investigative work should be comprehensive and not confined solely to subjects requested by the attorney/client.
- Testimony should always be honest and nonevasive.
- Data should not be tampered with in any way.

Activities that we consider to be professionally acceptable include

- Assisting the attorney in the preparation of questions and exhibits for direct examination
- Assisting the attorney in the preparation of questions for rival experts, both for depositions and for trial
- Sitting in and taking notes on rival expert testimony
- Participating in strategy sessions with the attorney

Many suggestions have been made within the engineering and geology professions to combat the present unacceptably high level of advocacy and unethical behavior. In 1989 an editorial featured in American Society of Civil Engineers magazine *Civil Engineering* was entitled "The Design Professions: Let's Regulate Expert Witnesses." The author (Friedlander, 1989) makes the point that lawyers can be disciplined for unethical behavior but that design professionals cannot. He believes that professional societies should police themselves by developing procedures to monitor the opinions of expert witnesses and to discipline those whose testimony is without merit. The author emphasizes "that even the most insincere opinions never rise to the level of perjury, and as long as two or more professionals offer conflicting opinions, neither side will be deemed to have adopted frivolous positions."

Specifically the author suggests that the professional societies form committees to review suspect opinions rendered by their members. He feels that aggrieved litigants or their lawyers would be willing to submit appropriate transcripts to the committee for review. If a committttee found an expert witness member to have rendered an opinion that it considered not to be in conformance with generally accepted professional standards, it could censure the member by:

- Issuing a public statement of censure that could be used to discredit the member at a future trial
- Suspending or revoking membership in the professional society
- Reporting the professional misconduct to the expert's state licensing board

If it could be successful there would be no better way to handle the problem than from within the professions. But we have serious reservations. For example, neither the American Medical Association nor the American Bar Association, which actually possess regulatory measures in their charters and bylaws, have ever been able to control the ethical behavior of their members. Criminal behavior, maybe—ethical behavior, no.

There is no reason to believe an engineering or geologic society would be able to exert much if any pressure on its members to make more honest expert witnesses of them. It would be extremely difficult to interpret what is in the mind of a witness from reading a trial transcript. One committee member might find an opinion dumb but sincere, while another might find it flagrantly dishonest. Cenorship on the basis of committee consensus might provoke a counter lawsuit in itself, which would be very difficult for the professional society to win. Besides, there are probably very few engineers and geologists who would be willing to render this kind of judgement against their peers unless the behavior bordered on criminality. Certainly this fact has always been true in the medical and legal professions. As far as state licensing boards are concerned, they would probably be very hesitant to consider taking action against anyone in an activity as nebulous as rendering an opinion in a lawsuit. The boards appear to have their hands full handling gross infractions such as practicing without a license. We are not suggesting that self-policing through societies should not be tried. We are simply very skeptical that it would succeed.

An intriguing alternative to policing by professional societies as a means of overcoming the advocacy problem among expert witnesses appeared in a 1988 issue of *Physics Today*. In a very eloquent plea for greater scientific objectivity in the courtroom, the author (Ahearn, 1988) suggested that the engineering and scientific societies should provide "friends of the court committees." The members, carefully chosen by the societies, would be extremely well qualified in their special fields and would be available to the courts either on a pro bono basis or would be paid by the courts. They would give expert witness testimony at the trial in addition to that given by the witnesses for each side. With this arrangement the experts for the plaintiffs and defendants could act as out-and-out advocates, with no restrictions other than those that are self-imposed to retain a semblance of credibility and self-respect. The advantages of having a highly qualified impartial "friend of the court" expert are:

- The judge and/or jury would be presented with an impartial third opinion, obviating the necessity of choosing between two diametrically opposite sets of opinions and trying to decide which experts seem most credible. Many times in awarding dollar damages a jury will merely arbitrarily split an award exactly down the middle between the plaintiffs' and the defendants' positions. The presence of the court's expert, in our opinion, would lead to fairer verdicts and better precedents.
- It could possibly reduce the number of frivolous suits since the plaintiff's attorney would not be able to rely on winning by overwhelming or descrediting the defendant's expert witness.
- The presence of the court's expert might moderate the other expert's testimony and keep it from becoming too outrageous.

Although we feel the idea of a court-employed expert has real merit, it probably also has limitations. For instance, its added cost might not be justified in smaller cases and its mere suggestion to a trial lawyers association would probably incur a tumultuous negative response.

Faced with the legal system as it exists today, what can the conscientious civil engineer or geologist do to minimize the risk of falling into the advocacy trap?

1. Choose one's forensic assignments carefully. If not already familiar with the case, we believe that it is wise to conduct a preliminary investigation of the salient facts before commitment. Most attorneys are willing to pay for such an investigation if it is done in a confidential manner. Depending on the findings, the attorney has several options:

- Retain the expert for a full investigation and as an expert witness if necessary
- Dismiss the expert if findings are adverse
- Settle the case at that point
  The advantages to the expert are:
- Payment for services regardless of findings
- Knowledge that, if retained, it will probably be a comfortable situation professionally
- Allows the expert to develop independent preliminary hypotheses
  The concept of a compensated preliminary report is especially applicable when attorney and expert do not know one another. It is not uncommon for an unscrupulous lawyer to refuse to pay for adverse findings.

2. Do not yield to an attorney's insistence on testifying from a prepared (and often rehearsed) script. It is an excellent idea to thoroughly discuss questions and answers with the attorney so that there will be no surprises at trial. However, in our opinion, if carried to the extreme of an actual script it

becomes unprofessional and, many times, counterproductive both scientifically and in credibility to the court.

3. Insist on performing a full and complete investigation of all facets of the disaster within the area of expertise.

4. Avoid conflicts of interest or even appearance of conflict. Before accepting an assignment it is important for an expert to be absolutely certain that none of the following situations exist:

- That there is no outside financial interest linking the expert to any of the parties involved in the case
- That there is no family relationship involved in the case, however remote
- That there has been no previous legal action in another matter that could be construed as presenting a conflict
- That there is no personal interest in the outcome of the case

5. Do not fall into the trap of allowing legal strategy to affect technical conclusions or opinions.

Example In a case in which we participated a geotechnical engineer with formidable credentials was employed to investigate the causes of earth settlement in a large multiresidential complex and subsequently to recommend mitigation measures. Substantial work was done to achieve this objective, including many deep soil borings. When the engineer submitted his findings and recommendations as part of an overall report, the intent was strictly to produce the best method of mitigation, untainted by legal input. One of the most important components of his investigation was to extrapolate the extent of anticipated future settlement. Based largely on his calculations, specific recommendations were made to the interested parties as to how the buildings should be repaired and future settlement damage minimized.

Before the mitigation measures could be approved and implemented, the attorneys from both sides swarmed in like a pack of hungry coyotes and took over. More geotechnical reports from other experts, most self-serving to their particular clients, were generated. Then, with the urging of his client's attorney, the engineer changed his original opinion quite drastically relative to the magnitude of future settlement. In all fairness to him, he may have been honestly and sincerely influenced in his change of opinion by the work of some of the other experts. At any rate, in our view, it was not his finest professional hour.

As in many of these ethical issues, there is a fine line to be drawn in the attorney/expert relationship. For instance, there is certainly nothing wrong with the attorney suggesting changes in an expert's report or testimony that would clarify or expand a particular matter. Interfering with the substance of the report is another matter.

Forensic work requires constant self-vigilance and soul-searching. At about the time an expert feels that he or she has seen everything in the way of pressures and ethical decision-making situations, something new will arise. Some mistakes and transgressions are bound to occur, particularly in the heat of legal battle. But it should be the aim of every engineer and geologist doing forensic work to strive for a high level of professionalism.

In theory a person should remain on the highest moral ground. In practice choices are often not that easy, as can be seen in the following hypothetic example: A geotechnical engineering expert is engaged to investigate the cause of excessive structural settlement in an area where there is deep fill. The engineer drills a half dozen test borings and performs all the appropriate test and calculations. Certain conclusions are then drawn regarding the cause and effect of the settlement, which are all highly favorable to the client's legal position. Just before trial testimony is to begin, the engineer discovers, accidentally and from an obscure source, logs of several previous borings in the area. The discovery, if disclosed, would be detrimental to the client's case. Should the engineer do nothing, hoping the opposition will not discover the logs? Or should the attorney be informed of the new evidence? And if the attorney's instructions are to forget having seen the logs, what then? It is a tough call, but the attorney should be informed and the expert should not agree to any tactic that could lead to perjury.

Frivolous Lawsuits

At least 25% of the lawsuits in which we have been employed could be classified as frivolous, in other words completely lacking in technical merit, and should never have gotten to court. Why did they proceed so far? Because enough of them were winners for the plaintiffs' attorneys to encourage others, armed with similar allegations, to follow suit. Why were some of these frivolous lawsuits successful? Because the plaintiffs' lawyers or experts were more skillful, better prepared, or more attuned to the judge's built-in bias than were counsel for the defendants. We used to believe naively that all judges, if not competent in environmental law, were at least scrupulously fair and evenhanded. We once scoffed at the notion there were "plaintiff's judges" and "defendant's judges." That is not the case anymore. Some judges not only possess certain biases but are influenced by them in their handling of particular cases, which is evident in their instructions to the jury and in their selection/classification of allowable exhibits.

At any rate, it seems apparent that successful environmentally related frivolous lawsuits breed more of the same. They constitute an enormous cost to society, not only because the public usually pays the price of the judgment but because they jam our already crowded courts, thereby unfairly delaying legitimate lawsuits. In view of the monetary rewards that can be achieved, there seems little that can be done within our present judicial system to stem the tide of frivolous suits. A beneficial change would be to force an unsuccessful plaintiff to pay all court costs (as is the case in Great Britain). Forensic experts could help the situation by not accepting plaintiff assignments that are obviously going to rely on radical departures from scientific honesty and integrity. Unfortunately there are too many civil engineers and geologists who either believe there is nothing professionally wrong with being an advocate or simply do not care as long as the fees are high enough.

Incompetent Professional Work

We are of the opinion that at least 90% of the cases involving landslides and other earth movement failures could have been avoided with competent civil engineering, geotechnical, and geologic input in conjunction with quality control throughout the project (from initial site analysis to completion of construction).

Civil engineers and geologists must realize that they have an assigned responsibility to function with competence and integrity as well as assuming the attendant legal liability for their work. Code requirements and enforcement may differ from one jurisdiction to another, but in recognizing that natural hazards are not constrained by political boundaries, courts are now tending to hold professionals accountable to a higher standard of care than the "standard practice" or "we met the code" concepts.

In many jurisdictions, codes and especially code enforcement may not meet current responsible professional standards; thus the civil, geologic, and geotechnical professionals cannot rely on code standards to immunize them or their clients from liability. The theory is that the level of professional work should equal that which a prudent person would meet considering current state of the art. Fraud and gross negligence are now being joined to the principle of simple negligence as grounds for legal action against professionals whose work proves to be erroneous, lacking in data, or inadequate. This is beginning to appear in some legal briefs.

There is a clear ethical and professional responsibility to maintain a high professional standard of care without regard to whether the codes within a particular jurisdiction are law or are adequately enforced. Further, one should not rely on codes as a standard inasmuch as upgrading of codes usually follows professional knowledge by about 10 yr (Fig. 11–1).

Politically active public advocacy groups such as property owners' associations have become commonplace, particularly in hillside areas plagued by fires and slope failures and in coastal areas subject to landslide and aggressive surf erosion. As the public advocacy groups, insurance companies, lawmakers, and plaintiffs' attorneys develop their own body of knowledge regarding natural hazards, the several professional disciplines involved in development within potentially hazardous areas have been compelled to share liability and cost of losses when failures occur. The act of God defense is no

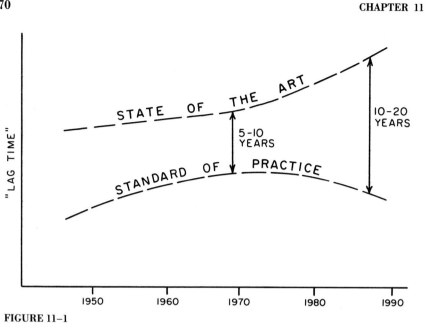

FIGURE 11–1
Estimated differential between standard of practice and state of the art.

longer considered tenable for most natural disaster cases since few natural hazards are unforeseeable to the responsible professional.

Codes and Regulations

We have concluded that, in general, government appears to be failing to enforce its own codes, regulations, and policies (exacerbated by the fact that, in many instances, the codes are out of date, have been weakened, and are inadequate). For example, flood damage, debris flow, and landslide damage along the western front of Utah's Wasatch Mountains in 1983 caused serious damage and monetary loss in connection with homes and public facilities, most of which were attributable to failure by government to properly enforce floodplain management regulations. A portion of the losses should be assigned to lack of technical knowledge and scientific/engineering experience, and another portion to the failure to use appropriately the limited expertise available. However, the greatest error by state and local governments is that of the failure to adopt and enforce codes and regulations. It is the authors' opinion that greater than 50% of the 1983 flood–debris flow–landslide damages and losses in Utah could have been prevented without impacting construction and economic growth or costs.

Land subsidence, landslides, and surficial slope or slump problems that have plagued some cities and counties in Southern California also are due to the failure, at least in part, to enforce adequately the existing grading codes and regulations at the permitting stage and further to utilize some form of inspection by government personnel or certification of work done by the professional to assure quality control.

In the City and County of San Diego, the rationale for not adequately enforcing codes and regulations has been:

1. It never rains in San Diego, so there is no reason to worry about subdrains, surface drains, or the quality of compaction.
2. We are enforcing the code or applying specifications that the consultants have told us are adequate.

Both or either of these are not realistic if one assumes that the quality of professional work will drop to the lowest level allowed by government.

The following quotes have been placed in the public record:

1. In response to questions by *The San Diego Union (September 9, 1982)* regarding what the City of San Diego has done to reduce the losses from landslide, one of the authors stated: "Local government in San Diego has done very little to correct the potential problems....[T]he City does not adequately enforce many of the minimum standards set by the State of California. . . . Local goverment in San Diego is very limply, weakly utilizing state-required codes. . . . The city could be a prime target in years to come if homeowners whose houses are ruined by a landslide decided to blame local government for allowing construction in an unstable area that would threaten public safety."
2. In the winter of 1981–1983 the San Francisco Bay area suffered damages from landslides, mud flows, and floods estimated to have been in excess of $300 million. The California State Board of Geologists and Geophysicists, as a regular function, conducted hearings and field trips related to these geologic hazards. Following the hearing and field trips, the board passed the following resolution at its next monthly meeting: The responsibility for the damages and losses from the January mud slides should be placed at the door of many of the local governments for:
   (a) failure to adequately enforce Chapter 70 of the Uniform Building Code;
   (b) failure to have licensed cerified engineering geologists on staff available via contract to review all geologic reports submitted to local government and to inspect geologic conditions periodically during grading and construction, in order to assure that the codes and standards are properly complied with.

The board was also critical of the State of California Department of Housing and Community Department for failure to police local government. In additon, the board was informed by legal staff that "recent decisions by the court have recognized the liability of local government for failure to enforce established statutes, codes and ordinances."Again, government was found, in part, responsible for the losses because of its failure to provide inspection to assure quality control.

Again, the most often used excuse or rationale is "we met the code," or "we did everything required by the city or county," or"the city or county accepted it." Thus government becomes, at least partly, the excuse as well as the cause. Some of the most successful plaintiffs' attorneys in California have been including local government as a defendant when damage due to poor quality engineering, geology, or construction caused losses. Some have even made local government the prime defendant; local governments have lost a few such cases and are currently facing many more suits at much higher dollar amounts.

3. In the June 1985 issue of *Civil Engineering,* a letter to the editor J. Haimowitz stated:

> Engineers sell their services to repeat customers who are experts at beating down prices in a market flooded with too many licensed engineers. Owners are asking engineers to bid for the privilege of playing chicken, to see who can do the least work and still get by.
>
> Market forces are making engineers choose between playing chicken and bankruptcy. This clash of corporate titans is causing a problem in the lower ranks of all organizations involved in the project. Construction contractors have no pride; the craftsman is gone.

The authors' exposure to the applied technology field over the past 40 yr supports the previous quote. Thus, if the client is applying pressure on the consultant, and if the client, consultant, and politicians are all applying pressure on the local, state, and federal government, elected officials, and employees, it seems apparent that goverment will usually allow the standard of practice and professional work to drop to a standard desired by the client.

In our experience of analyzing and preparing case histories, we have noticed another common denominator of failure or error. This is related to the often requested assignment by government or industry to look for a problem or to provide inspection and supervision but at the same time to remember that one is being paid *not* to find the problem or defect. Often a government agency issues a permit with a condition of approval that on-site inspection and supervision be provided and that design changes or mitigation be initiated if any unforeseen conditions are encountered. In this situation, the design and construction professionals are under pressure not to look or be replaced.

Educational Shortcomings

Part of the overall quality control problem that plagues projects involving civil engineers and geologists may lie in the education arena. It is apparent to the authors that most undergraduate programs have simply not kept pace with the demands of highly sophisticated planning, design, and construction procedures. In our view it is imperative that civil engineers and geologists not only be well versed in the fundamental technical aspects of their professions but be intellectually conversant with other professionals, including planners, environmentalists, administrators, and attorneys. Those that eventually attain positions of leadership must be able to comprehend the big picture (in other words, have peripheral vision and thought processes) in addition to having a thorough understanding of their field of specialization. Good graduate programs and continuing education in the form of seminars and special classes are more important than ever. Continuing education should be made a part of the licensing requirements for engineers and geologists.

Not so many years ago civil engineering was viewed as a low-tech profession in contrast to other branches of engineering. That this is no longer true must be recognized by civil engineering schools.

Within the field of geology there is a dearth of basic courses at most universities. The authors estimate that fewer than 5% of graduates in geologic sciences have taken or even had available to them courses in engineering geology, groundwater hydrogeology, marine geology (coastal processes), geomorphology, soil mechanics, hydrology, and other subjects pertinent to success in solving problems related to landslides and other types of land failures.

Future of Forensic Activity

What does the future hold for the practice of forensic civil engineering and geology? In the best of worlds one could hope that opportunities in the field would diminish as the result of more enlightened planning and design, stronger codes and regulations, tighter code enforcements, and more prudent operation and maintenance procedures—all accomplished with a thorough understanding of the potential legal consequences of a breakdown in any one of these areas. Unfortunately there is little reason for optimism at this time. The volume of forensic activity will probably expand until the citizenry realizes that it is paying a hugh price for the deficiencies of the system and demands that changes be made. In the meantime, forensic engineers and geologists should focus on upholding both personal dignity and the honor of their professions by performing in an objective manner using sound scientific principles.

GLOSSARY

Aggradation	The process of building up of a streambed by deposition.
Alluvial fan	A fan-shaped stream deposit built at the place where it issues from a narrow mountain canyon where the gradient of a stream is abruptly decreased and the flow spreads.
Aquiclude	A body of relatively impermeable rock that will absorb water slowly but will not transmit it fast enough to supply a well or spring.
Aquifer	A body of rock sufficiently permeable to conduct groundwater and to yield significant quantities of water to wells and springs.
Artesian	Pertaining to groundwater under sufficient hydrostatic pressure to rise above the aquifer containing it.
Artesian head	The hydrostatic head of the water in an aquifer.
Atterberg limits	In a sediment, the water-content boundaries between the semiliquid and plastic states (known as the liquid limits) and between the plastic and semisolid states (known as the plastic limit).
Bedding plane	In sedimentary or stratified rocks, the nearly planar divisions or surface that separates each successive layer or bed from the one above or below. A bedding plan is considered *adverse* if its dip is less than that of the surface slope.
Bed load	Larger particles of sand, gravel, and larger rock that slide, roll, or bounce along the streambed by traction or saltation.
Breccia	A coarse-grained clastic rock composed of large (greater than 2 mm) angular and broken rock fragments that are cemented together. Similar to conglomerate except that most of the fragments have sharp edges.
Bulking factor	A multiplier factor applied to peak clear water flow to account for the volume of sediment and debris in the flow as the result of a denuded watershed.
Common-law negligence	Negligence based on a "reasonable man" standard.

Corrasion The wearing away of the bed and banks of a stream by the cutting, scraping, scratching, and scouring effects of a sediment load carried by a stream.

Creep The slow, gradual, more or less continuous, nonrecoverable deformation sustained by soil and rock materials under gravitational body stresses.

Critical depth In an open channel, the depth of flow for which the discharge per foot of channel width is the greatest for a fixed head.

Declaration Unsworn statement or narrative of facts made by a party to the transaction or by one who has an interest in the existence of the facts recounted.

Dip The angle of inclination of the plane of stratification of rock units with the horizontal plane.

Electric log A geophysical log of an uncased part of a well or borehole that displays measurements of the in situ electrical properties of geologic formations encountered at various depths. It is useful for locating changes in composition, making local correlations, and indicating the nature and amount of fluids in the pores of the rock.

Evapotranspiration Combined loss of water from soils by evaporation and plant transpiration.

Factor of safety The stress at which failure is expected divided by the actual or design stress.

Freeboard The height between design/water level in a channel and the top of the channel.

Froude number (F) In an open channel the $\dfrac{\text{velocity}^2}{\text{length}} \times$ acceleration of gravity

Geomorphology The study of the classification, description, nature, origin, and development of landforms and their relationship to underlying structures. Also the history of geologic changes as recorded by these surface features.

HEC-1 A computer model program developed by the Hydrologic Engineering Center of the U.S. Army Corps of Engineers for computations of stream flows.

HEC-2 A hydraulic model program for computing water-surface profiles.

Hydrauger A (almost) horizontal perforated pipe to collect groundwater and transmit it by gravity. Installed by drilling a nearly horizontal boring and inserting a casing or pipe into the boring.

Hydraulic gradient An imaginary curve over a flowing pipeline that shows the levels to which the water would rise in open pipes. Also the slope of the surface of water flowing uniformly in an open conduit.

Hydraulic jump	A sudden rise in water surface that occurs in open channels when flow changes from fast flow (supercritical) to slow flow (subcritical). The jump is usually caused by obstacles to flow.
Hydraulic radius	The cross-sectional area of water flowing through a channel or conduit divided by the wetted perimeter.
Hydrograph	A graph with time as the abscissa and flow rate as the ordinate.
Hydrostatic head	The height of a vertical column of water, the weight of which, if of unit cross section, is equal to the hydrostatic pressure at a point.
Interrogatory	A set or series of written questions drawn up for the purpose of being propounded to a party in equity, a garnishee, or a witness whose testimony is taken on deposition; a series of formal written questions used in judicial examinations of a party or a witness. In taking evidence in depositions, the interrogatories are usually prepared and settled by counsel and reduced to writing in advance of the examination. Written questions propounded by one party and served on adversary, who must serve written answers under oath.
Inverse condemnation	A legal concept wherein property owners are entitled to just compensation if they can prove their property has been damaged as a result of a government project or action.
Isohyet	A line joining points of equal rainfall for a given period of time.
Kinematics	The science that treats motions considered in themselves, or apart from their causes. Measurements of displacement and direction.
Knickpoint	Any interruption or break in slope; especially a point of abrupt change in the longitudinal profile of a streambed, occurring where a new curve of erosion (graded to a new base level after a relative lowering of the former level) intersects an earlier curve.
Lithology	The description of rocks as based on the megascopic observation of hand specimens.
Mandatory duty	Concept wherein a statute creates a duty to perform an action.
Method of slices	A method of calculating the stability of an earth slope. The wedge of soil is divided into vertical slices of uniform width and the equilibrium of the whole wedge is calculated by summing the effects on all the slices.
Morphology	A scientific study of form and structure.
n factor	An empirical value that enters into any formula for flow in pipes or channels and expresses the friction or resistance to

flow due to the surface texture of the material in contact with the water.

Nonsuit — A term broadly applied to a variety of terminations of an action that do not adjudicate action issues on the merit.

Piezometer — An instrument used to measure fluid pressure and the elevation of water in the ground.

Pore water pressure — The stress transmitted through the fluid that fills the voids between particles of a soil or rock mass; for example, that part of the total normal stress in a saturated soil due to the presence of interstitial water.

Q — Maximum flow or peak discharge usually measured in cubic feet per second or cubic meters per second.

Return frequency — A statistical ratio representing the likelihood of a storm of a given magnitude and duration of occurring.

Reynolds number (R) — Dimensionless ratio of the inertial forces to the viscous forces $R = \dfrac{UL}{V}$ where U is the average velocity, L the length, and V the kinematic viscosity.

Roll waves — Intermittent surges on steep slopes that occur when the Froude number (F) is greater than 2.0 and the channel invert slope is greater than 12 divided by the Reynolds number.

Rule curve — A curve that shows the theoretic desired volume of water in a storage reservoir at any given time.

Sag vertical curve — A curve along the longitudinal gradient of a highway that is concave upward. The curve provides a gradual transition between two tangents with oppositely sloping gradients.

Slickenfibers — Elongated crystals generically related to the slickenside surface along a fault plane.

Slickensides — Polished and grooved surfaces on rock faces that have been subjected to faulting.

Standard project flood — A maximum design flow based on a combination of the most extreme conceivable meteorologic conditions for a particular watershed.

Stream braiding — The process of successive branching and rejoining of a stream channel, resulting in the formation of elongated islands and bars that split the channel into an intricate network of smaller interlacing channels.

Subrogation — Substitution of one person in the place of another with reference to a lawful claim, or right.

Summary judgement — Procedure whereby a judgement is rendered by the courts in advance of completion of trial, generally on a matter where there is no dispute as to facts.

Tectonic   Pertaining to the forces involved in the major structural or deformational features of the outer part of the earth such as folding and faulting.

Thalweg   The line connecting the lowest points along a streambed or valley; a longitudinal profile.

Tiltmeter   An instrument that measures slight changes in the tilt of the Earth's surface. It is used in volcanology and earthquake seismology as well as in motion related to landslides and other slope instability.

Time of concentration (t_c)   The time required for a particle of water falling on the most remote boundary of a watershed to reach the subject point.

Weir   An overflow structure across a channel to measure flow.

REFERENCES

TECHNICAL REFERENCES

Selected References and Recommended Reading

Ahearn, J.F., 1988, Addressing public concerns in science: Physics Today, v. 1, no. 9, pp. 36–42.

American Society of Civil Engineers, 1990, Quality in the constructed project, v. 1; Manuals and reports on engineering practice No. 73: ASCE, New York, New York.

American Society of Civil Engineers, 1985, Reducing failures of engineered facilities; Proceedings of a workshop sponsored by the National Science Foundation and the American Society of Civil Engineers, Clearwater Beach, Florida, January 7-9, 1985.

American Society of Civil Engineers, 1977, Sedimentation engineering: Manuals and Reports on Engineering Practice No. 54.

Arnow, T., 1984, Water-level and water-quality changes in Great Salt Lake, Utah, 1947-1983: U.S. Geological Survey Circular 913, p. 7.

Asquith, G.B., and Gibson, C.R., 1982, Basic well log analysis for geologists: American Association of Petroleum Geologists.

Brock, R.R., 1969, Development of roll-wave trains in open channels: W.M. Keck Laboratory of Hydraulics and Water Resources Report No. KH-R-16: California Institute of Technology, Pasadena, California.

Bushnell, M.M., 1980, Geology for planning: Imperial Beach 7 1/2 ft, quadrangle, San Diego, California: California Division of Mines and Geology Open-File Report 80-16.

California Department of Conservation, Resources Agency, 1981, Erosion and Sediment Control Handbook.

Campbell, R.H., 1975, Soil slips, debris flows, and rainstorms in the Santa Monica Mountains and vicinity, Southern California: U.S. Geological Survey Professional Paper 851.

Center for Philosophy and Public Policy, 1984, The zealous lawyer: Is winning the only thing?, *in* Report from the Center for Philosophy and Public Policy, v.4, no. 1: University of Maryland, College Park, Maryland, pp.1–5.

City of Rancho Palos Verdes, 1984, Feasibility of stabilizing Abalone Cove landslide: Report of the Technical Panel.

City of San Diego Planning Commission, 1977, Tia Juana River Valley plan.

Davis, J.D., 1980, Rare and unusual post-fire flood events experienced in Los Angeles County during 1978 and 1980, *in* Storms, floods, and debris flows in Southern California and Arizona 1978 and 1980:, Washington, D.C., National Research Council, National Academy Press, pp. 243–255.

DeBano, L.F., 1981, Water repellent soils: a state of the art: U.S. Forest Service, Pacific Southwest Forest and Range Experiment Station, Denver, Colorado, PSW-46.

Dib, A., and Grant, J.K., 1985, Legal handbook for architects, engineers and contractors: New York, Clark Boardman.

Ehlig, P.L. 1982, Mechanics of the Abalone Cove landslide including the role of ground water in landslide stability and a model for development of large landslides in the Palos Verdes Hills, *in* Geological Society of America, Landslides and landslide abatement, Palos Verdes Peninsula, Southern California: Published by the Association of Engineering Geologists Southern California Section, pp. 57–66.

Ellis, A.V., and Lee, C., 1919, Geology and groundwaters of the western part of San Diego Co., CA: U.S. Geological Survey Water Supply Paper 446, p.321.

Emery, K.O., 1960, The sea off Southern California, a modern habitat of petroleum: New York, John Wiley.

Esmilla, A.B., Phipps, M.B., and Slosson, J.E., 1987, Application of oil well technology and continuous coring to landslide investigation, *in* Geology of the Palos Verdes Peninsula and San Pedro Bay: Pacific Section of Society of Economic Paleontologists and Mineralogists and the Assocation of Petroleum Geologists, Los Angeles, California, pp. 2-7 – 2-11.

Fleming, R.W., and Schuster, R.L., 1985, Implications of the current wet cycle to landslides in Utah, *in* Delineation of landslide, flash flood, and debris flow hazards in Utah: Utah Water Research Laboratory General Series VWRL/G-85/3: Utah State University, Logan, Utah, pp.19–28.

Fleming, R.W., and Taylor, F.A., 1980, Estimating the costs of landslide damage in the United States: U.S. Geological Survey Circular 832.

Fleming, R.W., Varnes, D.J., and Schuster, R.L., 1979, Landslide hazards and their reduction: Journal of American Institute of Planners,. 45, no. 4, p.428–439.

Friedlander, M.C., 1989, The design professions: Let's regulate expert witnesses: Civil Engineering, April, p.6.

Hague, T.A., McClay, P.L., and Slosson, J.E., 1982, Geologic hazards and hillside properties, *in* Advanced real property series—earth movement: California Continuing Education of the Bar.

Haimowitz, J., 1985, Letter to the editor, Civil Engineering, June, 1985: American Society of Civil Engineers, pp. 33–34.

Hoggan, D.H., 1989, Computer assisted floodplain hydrology and hydraulics: New York, McGraw-Hill.

Jahns, R.H., and Vonder Linden, K., 1973, Space-time relationships of landsliding on the southerly side of the Palos Verdes Hills, California, *in* Geology, seismicity and environmental impact: Association of Engineering Geologists Special Publication, University Publishers, Los Angeles, California, pp. 123–138.

Kaliser, B.N., and Slosson, J.E., 1988, Snow-melt triggered debris flows affecting Utah's megalopolis: Hydraulic Engineering Proceedings of the 1988 National Conference of Hydraulics Division, American Society of Civil Engineers, Colorado Springs, Colorado, pp. 379–384.

Keaton, J.R., 1984, Field guide to selected landslide, debris flow and flood hazard areas, North-Central Utah: Utah State University Conference, Logan, Utah.

Krammes, J.S., 1965, Seasonal debris movement from steep mountainside slopes in Southern California: U.S. Department of Agriculture Paper 12, Miscellaneous Publication 970.

Krammes, J.S., 1960, Erosion from mountain side slopes after fire in Southern California: Pacific Southwest Forest and Range Experiment Station, U.S. Department of Agriculture, Research Note 171m.

Krohn, J.P., and Slosson, J.E., 1976, Landslide potential in the United States: California Geology, v. 29, no. 10, pp.224–231.

Kusler, J.A., 1985, Liability as a dilemma for local managers: Public Administration Review Special Issue.

Langbein, W.B., and Schumm, S.N., 1958, Yield of sediment in relation to mean annual precipitation: Transactions of American Geophysical Union, v. 39, no.6.

Legget, R.F., and Karrow, P.F., 1983, Handbook of geology in civil engineering: New York, McGraw-Hill.

Los Angeles County Flood Control District, 1982, Engineering methodology for mudflow analysis.

Los Angeles County Flood Control District, 1971 and additions, Hydrology manual.

Los Angeles County Flood Control District, undated, Design manual for debris dams and basins.

Martin, M.W., and Schinzinger, R., 1989, Ethics in engineering, 2nd ed. New York, McGraw-Hill.

Merriam, R., 1960, Portuguese Bend landslide, Palos Verdes Hills, California: Journal of Geology, v. 68, p. 140–152.

National Research Council, 1988, Estimating probabilities of extreme floods: Washington, D.C., National Academy Press.

Nilsen, T.H., Taylor, F.A., and Brabb, E.E., 1976, Recent landslides—an estimate of economic losses and correlation with slope, rainfall, and ancient landslide deposits: U.S. Geological Survey Bulletin 1398.

Ozment, J., 1983, The Thistle, Utah landslide: Denver & Rio Grande Western Railroad, Document No JLO 0811283.

Peterson, P.R., 1952, Geology of the Thistle Area, Utah: Brigham Young University, Master's thesis.

Poertner, H.G., 1988, Stormwater management in the United States: Bolingbrook, Illinols, Stormwater Consultants.

Poland, J.F., and Ireland, R.L., 1988, Land subsidence in the Santa Clara Valley, California, as of 1982: U.S. Geological Survey Professional Paper 497-F.

Rawson, R.R., 1957, Geology of the southern part of the Spanish Fork Peak Quadrangle, Utah: Brigham Young University, Master's Thesis, p. 3.

Rigby, J.K., 1962, Some geomorphic features of the southern Wasatch Mountains and adjacent areas, in Brigham Young University Geology Studies, v. 9, part 1, p. 81.

Rigby, J.K., 1968, Guide to the geology and scenery of Spanish Fork Canyon along U.S. Highways 50 and 6 through the Southern Wasatch Mountains, Utah: Brigham Young University Geology Studies, v. 15, part 3, p. 17.

Rigby, J.K., 1976, Field guide, Northern Colorado Plateau, Kendall/Hunt Publishing Company, Dubuque, Iowa, p. 5.

Ritter, J.R. and Brown, W.M. III, 1971, Turbidity and suspended-sediment transport in the Russian River Basin, California: U.S. Geological Survey Open-File Report, Water Resources Division.

San Diego County Flood Control District, 1980, Storm report, February 1980.

Schuster, R.L., and Krizek, R.J., 1978, Landslides analysis and control: Transportation Research Board Special Report 176, Washington, D.C., National Academy of Sciences.

Shroder, J.F., 1971, Landslides of Utah: Utah Geological and Mineral Survey Bulletin 90, p.27.

Sigma Xi, 1991, Honor in science: Research Triangle Park, North Carolina.

Sumison, O.B., 1983, Thistle... focus on disaster: Springville, Utah, Art City Publishing.

Troxell, H.L., 1954, Hydrology of the San Bernardino and Eastern San Gabriel Mountains, California: U.S. Geological Survey Hydrologic Investigation Atlas HA-1.

U.S. Army Corps of Engineers, Hydrologic Engineering Center, 1987, HEC-1 flood hydrograph package, users manual: Davis, California.

U.S. Army Corps of Engineers, Hydrologic Engineering Center, 1982, HEC-2 water surface profiles, users manual: Davis, California.

U.S. Army Corps of Engineers, 1965, Report on flood control, Tia Juana River Basin: Table of peak discharge data for floods on the Tia Juana River near Nestor, California for the 79-year period, 1884–1963.

U.S. Army Engineer District, Los Angeles, California, 1976, Final environmental statement, Tijuana River flood control project, San Diego County, California: prepared for United States Section, International Boundary and Water Commission, USS, IBWC-EIS-76-1.

U.S. Department of Agriculture, Soil Conservation Service, 1986, Urban hydrology for small watersheds.

U.S. Department of Housing and Urban Development, 1977, Flood insurance study, City of Imperial Beach.

U.S. Geological Survey, 1986, Estimating the costs of landslide damage in the United States: Circular 832.

Varnes, D.J., 1978, Slope movement types and processes, in Landslides: Analysis and Control; Transportation Research Board Special Report 176, edited by R.J. Schuster and R.J. Krizek: National Academy of Sciences, Washington, D.C., p. 11.

Wells, W.G., II, 1981, Some effects of brush fires on erosion processes in coastal California, in Erosion and sediment transport in Pacific Rim steeplands: Christchurch, New Zealand, International Association of Hydrologic Sciences Publication 132.

Wells, W.G., II, and Brown, W.M., III, 1982, Effects of fire on sedimentation processes, in Environmental qualitites lab report 17-D: Pasadena, California, California Institute of Technology.

Wiggins, J.H., 1978, Building losses from natural hazards: yesterday, today and tomorrow: Washington, D.C., National Science Foundation Grant No. ENV-77-08425.

Williams, C.P., 1934, Foundation treatment of Rodriguez Dam: American Society of Civil Engineers Paper 1863.

Witkind, I.J., and Page, W.R., 1983, Geologic map of the Thistle area, Utah County, Utah: Utah Geological and Mineral Survey Map 69.

Woodring, W.P., Bramlette, M.N., and Kew, W.S.W., 1946, Geology and paleontology of Palos Verdes Hills, California: U.S. Geological Survey Professional Paper 207.

Yerkes, R.F. and Campbell, R.H., 1980, Geologic map of east-central Santa Monica Mountains, Los Angeles County, California: U.S. Geological Survey Map I-1146, Scale 1:24,000.

PUBLISHED REPORTS AND PAPERS BY THE AUTHORS

Graham, A., Shuirman, G., Slosson, J.E., and Yoakum, D.D., 1983, Hidden Springs flood: Proceedings of the D.B. Simons Symposium on Erosion and Sedimentation: Fort Collins, Colorado, Simons, Li and Associates pp. 1038–1053.

Shuirman, G., and Slosson, J.E., 1986, Preparing to be an expert witness: Proceedings of Association of State Floodplain Managers Symposium on Western State High Risk Flood Areas, Las Vegas, Nevada, pp. 153–159.

Shuirman, G., Slosson, J.E., and Yoakum, D.D., 1984, Relationship of fire/flood, *in* Delineation of landslide, flash flood, and debris flow hazards in Utah; Utah Water Research Laboratory General Series UWRL/G-85/3: Utah State University, Logan, Utah, pp. 178–194.

Slosson, J.E., 1986, Avoiding liability and other legal problems: Proceedings of Association of State Flood Plain Managers Symposium on Western State High Risk Flood Areas, Las Vegas, Nevada, pp.145–147.

Slosson, J.E., 1985, Development of geologic/seismic regulations and criteria: Bulletin of the Association of Engineering Geologists, v. 22, no. 1, pp.11–23.

Slosson, J.E., 1985, Encouraging government mitigation: The forensic geologist/ expert witness perspective: Proceedings of the American Bar Association International Symposium, Miami, Florida.

Slosson, J.E., 1984, Genesis and evolution of guidelines for geologic reports: Bulletin of the Association of Engineering Geologists, v. 21, no.3, pp.295–316.

Slosson, J.E., 1983, *Sprecher vs. Adamson Companies:* A critique of the Supreme Court decision: California Continuing Education of the Bar, v. 6, no. 6, pp 117–120.

Slosson, J.E., 1979, Recognition of landslides, *in* Geomorphic applications in engineering geology—a state-of-the-art short course: Office of Continuing Education, California State University at Los Angeles, pp.113–139.

Slosson, J.E., 1969, The role of engineering geology in urban planning: Colorado Geological Survey Special Paper 1, pp. 8–15.

Slosson, J.E. and Havens, G.W., 1987, Mitigation rather than litigation of the Abalone Cove landslide, *in* Geology of the Palos Verdes Peninsula and San Pedro Bay: Pacific Section of the Society of Economic Paleontologist and Mineralogists and the American Association of Petroleum Geologists, Los Angeles, California, pp. 2-3 – 2-16.

Slosson, J.E., and Havens, G.W., 1985, Government appears to be failing in enforcement: Proceedings of Natural Hazards Research Workshop, University of Colorado, Boulder.

Slosson, J.E., Havens, G.W., Shuirman, G., and Slosson, T.L., 1987, Harrison Canyon debris flows of 1980, *in* Landslides in a semi-arid environment with emphasis on the inland valleys of Southern California: Riverside, California, (Sadler, P.M., and Morton, D.M., eds.) The Inland Geological Society. pp. 285–297.

Slosson, J.E., and Krohn, J.P., 1980, Southern California landslide of 1978 and 1980, *in* Storms, Floods, and Debris Flows in Southern California and Arizona, 1978–1980: Washington, D.C., National Research Council, National Academy Press.

Slosson, J.E., and Krohn, J.P., 1979, AEG building code review, mudflow/debris flow damage, February 1978 storm, Los Angeles area: California Geology, v. 32, no. 1, pp. 8–11.

Slosson, J.E., and Krohn J.P., 1977, Effective building codes: California Geology, v. 30, no. 6, pp.136–139.

Slosson, J.E., MacArthur, R.C., and Shuirman, G., 1987, Legal misuse of urban hydrology concepts and regulations for rural areas, *in* Engineering hydrology: American Society Association of Civil Engineers, (Feldman, A.D., ed.) pp. 714–719.

Slosson, J.E., and Slosson, T.L., 1988, Pinto Wash alluvial fan litigation: *in* Hydraulic Engineering Proceedings of the 1988 National Conference of Hydraulics Division of American Association of Civil Engineers, Colorado Springs, Colorado, pp. 49–54.

Slosson, J.E., and Slosson, N.J., 1987, Agency control and damage reduction, *in* Landslides in a semi-arid environment with emphasis on the inland valleys of Southern California: Riverside, California, (Sadler, P.M., and Morton, D.M., eds.) The Inland Geological Society. pp.79–84.

Slosson, J.E., Shuirman, G., and Yoakum, D.D., 1986, Responsibility/liability related to mudflows: Water Forum '86: World Water Issues in Evolution, American Society Association of Civil Engineers, pp.235–243.

Slosson, J.E., Yoakum, D.D., and Shuirman, G., 1986, Thistle, Utah landslide: Could it have been prevented?: Proceedings of the 22nd Symposium on Engineering Geology and Soils Engineering, Idaho Department of Transportation, Boise, Idaho, pp. 281–303.

Yoakum, D.D. and Slosson, J.E., 1986, Preparing and conducting litigation dealing with arid flood problems, *in* Delineation of landslide, flash flood, and debris flow hazards in Utah: Utah Water Research Laboratory General Series UWRL/G-85/3: Utah State University, Logan, Utah, pp. 160–162.

LEGAL REFERENCES

Legal References Pertaining to Case Histories

Chapter 3
Hansch, et al. vs. County of Los Angeles, et al., 1988, California Court of Appeals, Second Appellate District, Division Two, Case No. B022074 (Superior Court for Los Angeles Case No. WEC 86007).

Chapter 4
Berrett et al. vs. Denver Rio Grande Western Railroad Company., 1986, 4th District Court for Utah County, CV-616 on appeal, Utah Court of Appeals Case No. 910215-CA.

Denver and Rio Grande Western Railroad Company vs. Utah Railway Company, et al., 1983, 4th District Court for Utah County, Civil No. C83-1130 J (settled).

Chapter 5
Albers vs County of Los Angeles, 1965, 62 Cal 2d, 250 (Portuguese Bend landslide).

Horan vs. County of Los Angeles, 1984, California Superior Court for Los Angeles County Case No. C294752 (settled).

Chapter 6

General Telephone Company vs. The City of San Diego, 1983, California Superior Court for San Diego County, Case No. 463619.

Chapter 7

Brisk vs. Safeco Insurance Co., 1981, United States District Court, Central District of California, Case No. CV81 1478LEW (TX) (settled).

Chapter 8

First English Evangelical Lutheran Church of Glendale vs. County of Los Angeles, 1984, Superior Court for Los Angeles County, Case No. C273634 (nonsuit).

Hidden Springs vs. Los Angeles County Flood Control District, 1984, California Superior Court for Los Angeles County, Case No. NEC 26146.

Chapter 9

Aalso, et al. vs. Santa Clara Valley Water District, et al., 1988, California Superior Court for Santa Clara County, Case No. 539738 (settled).

Chapter 10

Dry Creek Valley Association vs. Soiland Company, 1978, California Superior Court for Sonoma County, Case No. 84296 (settled).

Related Legal References

Ward Concrete Products Co. vs. Los Angeles County Flood Control District, 1957, 149 Cal. App. 2d 840, 847. (Flood Control District held liable for flooding that occurred as a result of their emplacement of a debris-collecting grate on storm drain, which impeded the flow of storm water causing overflow.)

Hughes vs. Potomac Insurance Co., 1962, 199 C.A. 2d 239. (Landslide caused by rain and groundwater was judged not subject to insurance policy exclusion for "flood waters" or "overflow of streams.")

Sabella vs. Wisler, 1963, 59 Cal 2d 21. (Where there are several different causes for fill failure, the efficient cause—the one that sets the others in motion—is the cause to which the loss is to be attributed.)

Connor vs. Conejo Valley Development Co., 1968, 69 A.C. 887, 447 P. 2d 609. (Construction lender could be liable in negligence for defects created by expansive soil.)

Avner & Avner vs. Longridge Estates, 1969, Cal App. 2d 272. (Developer was held strictly liable under tort for fill failure caused by inadequate drainage.)

Shaeffer vs. State of California, 1972, 22 Cal. App. 3d 1017. (If flood control improvements do not subject a landowner's property to any additional flooding than would have occurred absent the improvement, the state incurs no liability.)

Tri-Chem, Inc. vs. County of Los Angeles, 1976, 60 Cal. App. 3d 306. (For plaintiff to recover on inverse condemnation, he must establish that flood control system did more harm than good.)

Ducey vs. Argo Sales and State of California, 1979, 25 Cal 3d 707, Calif. Supreme

Court S.F. No. 23931. (If a government agency is adequately warned of a hazard, it has an obligation to mitigate.)

Del Mar Beach Club vs. Imperial Contracting Co., 1981, Calif. Ct. of Appeals, Sept 24, 1981. (Government agency issuing a waiver becomes responsible for damage.)

Sprecher vs. Adamson Companies, 1981, LA 31394 Superior Ct. No. WEC 52924. (Uphill landowner can be held responsible for landslide damage to downhill property even if uphill property is in natural state.)

Baldwin vs. Carleton Santee Corp, 1982, Calif. Superior Ct. for San Diego, 427 722. (Court found for homeowners against developer when ancient slide under tract moved. Developers had knowledge of ancient slide but found consultant who ignored it.)

Safeco Insurance Co. of America vs. Guyton,1982, 692 F 2d 551. (Flood damage was presumably excluded from insurance policy. However, policyholder was able to collect because of the negligence of those who constructed certain flood control facilities.)

Hildre vs. Imperial Irrigation District, 1983, 146 Cal. App. 3d 902. (State was liable in inverse condemnation because land was flooded by waters due to construction of an undersized culvert.)

Premier Insurance Co. vs. Welch, 1983, 140 Cal. App. 3d 370. (House on fill was destroyed by a slide caused partly by rain—excluded from policy—but also partly by a damaged subdrain. Subdrain was a concurrent proximate cause, and therefore policy exclusion did not apply.)

Valley Circle Estates vs. VTN Consolidated, Inc., 1983, 189 Cal Rptr 871. (Ten-year statute of limitations does not apply to subcontractor guilty of negligent land development because he could not otherwise be reached in a direct action.)

Sheffet vs. County of Los Angeles, 1984, 3 CA. 3d 720. (County liable for damages because new subdivision roads altered surface drainage, allowing water and mud to flow on plaintiff's property).

Salton Bay Marina vs. Imperial Irrigation District, 1985, C.A. 4th Civ 26949. (Waivers of liability for flooding required of irrigation district before obtaining building permit were ruled improper. District was found liable for flood damages by inverse condemnation.)

Rooney vs. Union Pacific Railroad, 1986, U.S. Dist. Ct., Wyoming. (Act of God defense was judged not applicable in flooding case even though drainage structure was adequate for 100-yr flood and was actually damaged by 500-yr frequency flood.)

Ektelon vs. City of San Diego, 1988, Ct. of App., 4th A D 1d, D002575. (If upstream owner modifies streamflow, he must consider downstream consequences.)

INDEX